Principles of Environmental Toxicology

Principles of Environmental Toxicology

IAN C. SHAW
University of Central Lancashire, Preston

and

JOHN CHADWICK
Head of Environmental Risk Assessment Section,
Health and Safety Executive

UK	Taylor & Francis Ltd, 1 Gunpowder Square, London EC4A 3DE
USA	Taylor & Francis Inc., 1900 Frost Road, Suite 101, Bristol, PA 19007-1598

Copyright © Taylor & Francis Ltd 1998

British Library Cataloguing-in-Publication Data

A catalogue record for this book is available from the British Library.

ISBN 0-7484-0355-8 HB
ISBN 0-7484-0356-6 PB

Library of Congress Cataloging-Publication-Data are available

Cover design by Amanda Barragry

Illustrations by Stephen Cookson and Ian C. Shaw

Typeset in Times 10/12pt by MHL Typesetting Ltd, Coventry

Printed by TJ International Ltd, Padstow, UK

DEDICATION

To my Mother and Father

I.C.S.

To Cath, Rebecca and Eleanor

J.C.

Contents

Acknowledgements

This book grew out of the MSc in Environmental Toxicology that we run at the University of Central Lancashire, Preston, UK. We are indebted to the course's lecturers for ideas and stimulation, in particular Arthur Lally of the Central Veterinary Laboratory, Weybridge, UK, and Dr Jenny Woodhouse of the Centre for Toxicology, University of Central Lancashire, UK, for their invaluable help in the preparation of Chapter 6, and Dr Alexis Holden, also of the Centre for Toxicology, and Dr Peter Wierden of the Chemistry Department, University of Central Lancashire, for their help with Chapter 4.

We thank Audrey Shaw and Joanne Knight for help with the typing, Dr Cath Chadwick for typing and proofreading, Stephen Cookson for drawing the molecular diagrams and Dr Paul Fitzmaurice for solving our numerous word processing problems.

Finally, we thank Janie Curtis of Taylor & Francis Ltd for commissioning the text before her departure for pastures new and Elaine Stott for her incredible patience when we missed every deadline.

Acknowledgements

1

Historical Review of Human Impact on the Environment

> This chapter covers the evolution of pollution from the time when humans lit the first fire up to the present time when air pollution in our cities is beginning to affect human health.
>
> Major pollutants (SO_2, NO_x, sewage) and agrochemicals (OCs and OPs) are discussed.

1.1 The Development of Pollution

In recent years humans have become more aware of their environment. This probably relates to the realisation that if we continue depositing our waste products at our current rate the environment, our world as we know it, will have a finite lifetime. The worry is, however, that the definition of finite might be measured on a scale that humans can relate to directly. It is for this fundamental philosophical (and selfish) reason that we are becoming interested in the effects of the chemicals that we use upon our world. This is the study of environmental toxicology. As with any science, what things do represents the fundamentals of understanding. In the case of environmental toxicology the action of chemicals upon environmental systems (ecosystems) forms the basis of the science. One hopes that having some, albeit very small, understanding of the effects of exogenous chemicals upon the inhabitants of an ecosystem might permit our predicting and perhaps preventing their deleterious effects in the future.

Generally the evolution of a scientific subject is slow, commencing with investigations of fundamental mechanisms, followed by an understanding of these mechanisms and moving to prediction and global understanding. Environmental toxicology is quite different. It has evolved on a fast track scheme, being driven by worries about the environmental devastation caused by human activity, fired by the green lobby and forced by legislation. For this reason we are being cajoled into predicting and acting before we understand the fundamentals of the subject.

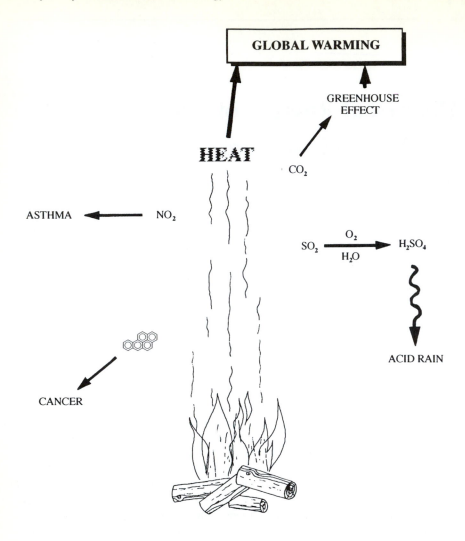

Figure 1.1   Generation of pollutants from fire.

Since humans evolved into a recognisable species about a million years ago their activity has modified the environment. Indeed it is a trait of humans that they not only evolved in a Darwinian manner, but also changed their environment to suit their needs. For example, in order to live in adverse climates people began to wear warm furs (these evolved into more complex clothes), they built homes and, very much later, dramatically extended their geographical distribution by developing air conditioning and central heating systems. Keeping warm in cold climates was an important step in human evolution. It was also a fundamental step in the acceleration of pollution. Fire produces carbon dioxide and other oxides (e.g. nitrogen dioxide, NO_2) (see Figure 1.1) which are important environmental

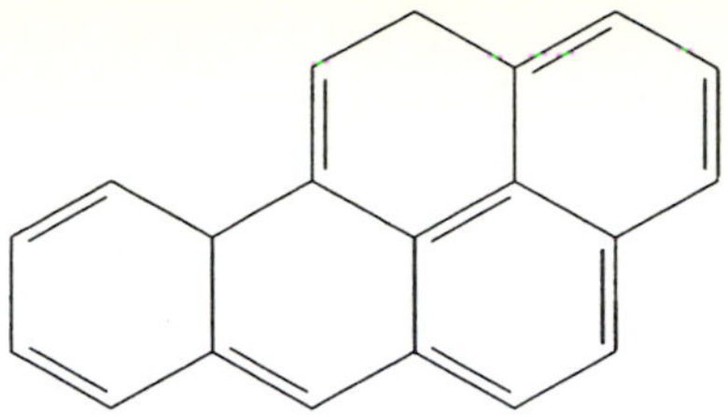

Figure 1.2 Molecular structure of benzo[a]pyrene, a carcinogen produced by combustion of carbonaceous materials.

pollutants. Both are dealt with in later chapters. The more esoteric product of the complex chemistry that occurs at the high temperatures within a fire are capable of significantly modifying the genetic material of both animals and plants. For example, benzo[a]pyrene, a polycyclic hydrocarbon (see Figure 1.2), modifies DNA resulting in mistranscription. In humans and higher animals this is likely to manifest as an uncontrolled division of the modified cell (cancer), whereas in lower organisms it is likely to result in genetically modified daughters. This modification, or mutation, might be beneficial or perhaps detrimental. It might give the organism an environmental advantage, weaken it or result in its death. This, of course, is the basis of evolution. Purists would say that fire is natural and therefore the whole process is biologically acceptable. This may well be so, but as the human population has grown and its dependence upon fire increased it is clear that the word *natural* might not be the most felicitous choice!

It is therefore clear that humans have, since time immemoriam, polluted their environment and that this pollution adversely affected their co-inhabitants. Being philosophical, one might conclude that this adverse effect upon co-inhabitants of the environment is all part of evolutionary supremacy. That may well be the case, but we are now in a situation with some 4.5 billion human inhabitants of the earth in which the adverse effects upon the environment may well limit the further population growth of humans themselves or, even worse, force them into decline. This concept, of course, is nothing new in biology. We are well aware of the simplest organisms growing initially uncontrollably (log growth phase) (see Figure 1.3) and then reaching a peak of growth rate (the point at which their environment can only just support, in terms of nutrients, the population), followed by a brief plateau and then decline and eventual extinction. A fatalistic view is that people are simply following this profile and there is absolutely nothing that we can do about it. This very pessimistic view would certainly be the case if we were a non-thinking *Daphnia*, but people can think and plan and act to prevent pollution being the force which pushes them into decline. Optimistically, the fact that humanity's future depends upon protecting its environment will have a very positive knock-on effect to most other living organisms. At last we see the need to protect our environment.

The concept that an increasing human population is responsible for environmental decline is not entirely fair. Increasing population size in human

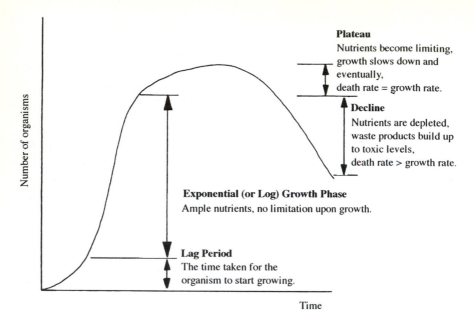

Plateau
Nutrients become limiting, growth slows down and eventually, death rate = growth rate.

Decline
Nutrients are depleted, waste products build up to toxic levels, death rate > growth rate.

Exponential (or Log) Growth Phase
Ample nutrients, no limitation upon growth.

Lag Period
The time taken for the organism to start growing.

Number of organisms

Time

Figure 1.3 Graphical representation of the growth phases of a bacterial liquid culture. This equally well represents the population kinetics of any other species, including humans.

terms ran hand-in-hand with increasing knowledge and there came a point when humans became aware that some of their activities damaged either themselves directly or their environment. There are several examples of this realisation in Greco-Roman cultures (1100 BC–AD 565). The Greeks (560–333 BC) began intensively farming their crops. They grew fields of grasses and caged and fenced animals for food production. In order to get the best out of their land they used the manure from their livestock to fertilise their crops. This is one of the earliest examples of the application of nitrates to crops and was the precursor to our current problems with nitrate excess in water courses (eutrophication). We return to these issues later.

It is clear then that human occupation of the planet goes hand-in-hand with pollution; however, the level of pollution and the effect that this has had on the fine equilibrium of nature is what we must consider. Early human effects upon the environment were small and in most cases local. In general these local ecological effects were reversible. For example, the small amount of nitrate that the Greeks put on their fields could be washed away by rain and diluted by rivers to such a great amount that it would have no perceptible effect on the biosphere. This *status quo* was perhaps the case until population intensities became great. For example, the Aztecs developed a highly populated society which depended upon the potato as a dietary staple, the massive monocultures which resulted decimated the ecology of the region resulting in denutrification of

4

the land and reducing crop yields. This was perhaps one of the reasons that the Aztec civilisation collapsed.

The most sinister date on the calendar of pollution is the commencement of the industrial revolution in the early 1800s. The industrial revolution marked the commencement of mass production, of mechanisation and (perhaps) of over-production. This enormous increase in productivity meant waste. Factories sprang up in areas close to important elements of infrastructure (e.g. canals for the transport of raw materials), waste products were discharged into the immediate environs (e.g. the canals) without any thought at all. Soon industrial connurbations were flourishing. They were financially rich areas employing many people. Such developments pushed out wildlife. The pollution emanating from these industrial cities (e.g. Birmingham, UK) devastated the environment for miles around. So great was this devastation and over so long a time period that some plants and animals adapted in such a way as to give them an evolutionary advantage.

A classic example of this is industrial melanism. Coal was used to fire the blast furnaces, to heat the water to provide the steam to drive pistons and to heat the large number of homes that were essential to house the workers around the edges of industrial areas. Among other (perhaps more damaging) pollutants arising from coal burning was fine particulate carbon. This blew in the wind and coated buildings and trees (and people's lungs) with a black layer. Over a period of time the colour of trees on the outskirts of cities changed from mottled browns and greens to ubiquitous black. Animals who normally hid on their bark or among their branches had evolved over millions of years to resemble the brown/green mottled background. Suddenly the background had changed and they were easily picked out by predators. Moth populations were quick to adapt to this change in their environment and the melanic (dark) forms were selected for and became predominant. The melanic form blended in with their new background so making them less visible to their predators and allowing them to maintain their population. This is an example of amazingly rapid natural selection. It is perhaps the only example of natural selection which could be witnessed in the lifetime of humans. The most studied moth, in this respect, is the peppered moth (*Biston betularia*). It lives on the silver and grey bark of birch (*Betula pendula*) trees and originally was mottled silver, white and grey itself (see Figure 1.4) in order to give it near perfect camouflage. Airborne carbon rapidly turned the silver bark black and made the moths visible to their predators. There was a rare dark (melanic) form of the moth which had resulted from an earlier mutation; this form was almost invisible on the newly darkened bark and so was selected into the population very rapidly indeed. Soon the melanic peppered moth became the most common form in industrial areas. In genetic terms the gene for melanism is dominant and so even the heterozygotes were dark; this of course speeded up the selection process significantly (see Figure 1.5).

The predominance of the melanic peppered moth in and around industrial areas was demonstrated by Kettlewell in 1973 when he reported the results of a

5

A

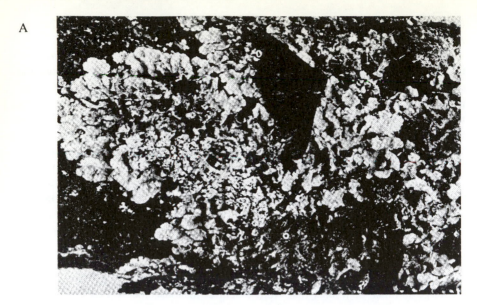

B

Figure 1.4 The peppered and melanic forms of *Biston betularia* on lichen-covered bark (A) and bark with carbon deposits (B). This shows clearly how well the melanic form is camouflaged on dark bark and the peppered form on lichen-covered bark. Reproduced from *Biology – a Functional Approach* by M.B.V. Roberts, with kind permission of Thomas Nelson Ltd, Sunbury-on-Thames, UK.

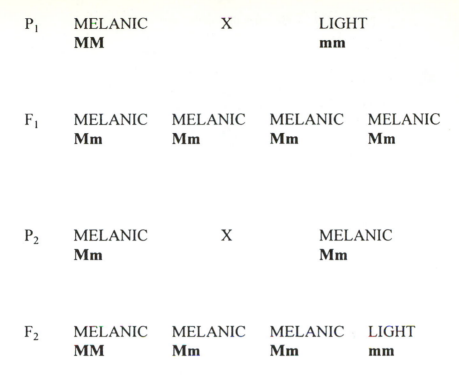

P₁   MELANIC   X   LIGHT

MM   **mm**

F₁   MELANIC   MELANIC   MELANIC   MELANIC

    Mm   **Mm**   **Mm**   **Mm**

P₂   MELANIC   X   MELANIC

    Mm   **Mm**

F₂   MELANIC   MELANIC   MELANIC   LIGHT

    MM   **Mm**   **Mm**   **mm**

Figure 1.5 Genetics of inheritance of melanism in the peppered moth (*Biston betularia*), showing how the dominance of the gene for melanism (**M**) over the gene for lightness (**m**) results in the melanic form predominating very quickly indeed. This, coupled with the susceptibility of the light form to being eaten by predators, meant that the melanic form quickly became the only phenotype found in industrial areas.

survey from 1952 to 1970 (see Figure 1.6). It can be seen clearly that there are more light forms of the moth in the rural areas of Devon, north Wales and highland Scotland than the industrial belt between London and Birmingham and the industrial areas between northwest and northeast England. From this one example it is clear that animals (and plants) will adapt to pollution; this example, however, is a success story. How many species have disappeared because of the deleterious effects of pollution? Take the salmon for example; it was once common in many rivers, including the River Thames which flows through London. The presence of a salmon in the Thames is now headline news, and all because of the chemicals that we have poured into the river from the industry along its banks. There are of course many other examples of species disappearing from environments as urbanisation or industrialisation occurs; gone are the days that you could land a trout in the Hudson River as it flows out of New York in the USA.

    The effects of carbon on our buildings and trees are obvious. Indeed an industry has built up around its removal – a trip to the beautiful city of Bath in the west of

Genetics of populations

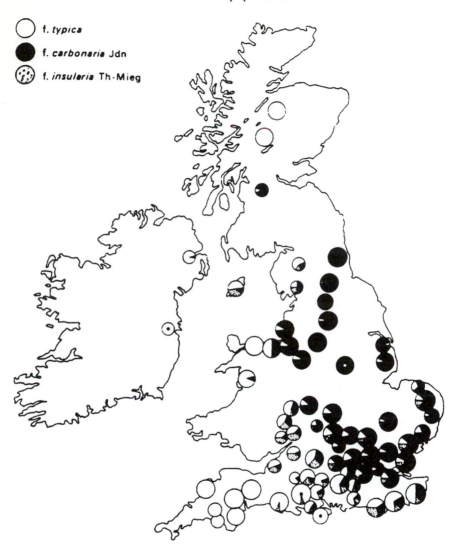

Figure 1.6 Distribution of the peppered and melanic forms of *Biston betularia* in Britain showing that the melanic form was commonest in and around industrial areas. From H. B. D. Kettlewell (1973), *The Evolution of Melanism with Special Reference to Industrial Melanism in the Lepidoptera*, Clarendon Press, Oxford.

England might be rather disappointing because so many buildings are covered with tarpaulin and plastic sheeting during the removal of this unsightly blackness. But what about the pollutants that we cannot see?

It is possible to trace back through many thousands of years the emission of invisible toxic gases. Combustion produces CO_2, NO_2 and SO_2. All are toxic by very different mechanisms. As combustion was synonymous with the development of industry, the industrial revolution resulted in an enormous increase in the emissions of the trio of environmentally toxic gases.

1.2   The Major Pollutants

1.2.1   *Sulphur Dioxide*

SO_2 is an increasingly important pollutant found in fossil fuel smoke and exhaust gases; in Britain about 4×10^6 tons of sulphur oxides were released into the atmosphere in 1986 compared with only 0.25×10^6 tons in 1800 (see Table 1.1). This makes SO_2 one of the major global pollutants of the twentieth century. This is the first mention of a global pollutant: the globality of SO_2 is determined by its presence in the air and the fact that the earth's prevailing winds carry gaseous pollutants across continents and seas. We return to this a little later because the transfer of pollution across territorial boundaries is a major political issue in the 1990s. Evidence of the effects of SO_2 can be seen by looking at lichens which are extremely sensitive to SO_2 pollution; in areas downwind of industrialisation there has been a marked decline in the prevalence of lichens on trees and buildings. Lichens are an excellent example of marker species and are very useful as an indicator of pollution.

It is important first to consider the mechanism of toxicity of SO_2. Wind blows the emission of SO_2 from a chimney stack hundreds of feet up into the

Table 1.1   Emissions of SO_2 into the atmosphere in Britain between 1800 and 1986

Year	SO_2 emissions (million tons)
1800	0.25
1870	2.50
1900	4.00
1939	4.00
1951	5.00
1965	5.80
1968	6.00
1974	5.00
1977	5.00
1986	3.70

Note the beginnings of a decline in 1986 due to increased environmental awareness and the introduction of legislation to clean up the air (Clean Air Act, 1956) 30 years earlier. Data from Clapp, B. W., 1994, *An Environmental History of Britain*, Harlow, UK: Longman.

atmosphere; from here it might be carried many hundreds of miles by air currents where it will be in contact with moisture and oxygen in the air. The SO_2 dissolves in the airborne water and is very rapidly oxidised by oxygen to form sulphuric acid:

$$SO_2 + H_2O + 1/2O_2 \rightarrow H_2SO_4$$

The H_2SO_4-containing rain falls as acid rain and results in the acidification of the land and waterways.

There is an important political issue here. Industrialised countries such as Britain pump many tens of thousands of tons of SO_2 into the atmosphere each year; however, very little of the resulting acid rain falls on their own land. Continuing with the British example, the prevailing westerly winds carry the H_2SO_4-bearing clouds to Scandinavia where, on meeting the mountains, they precipitate the acidity over Norway and Sweden and so change the natural pH of many of their lakes from slightly basic to acid. The effects of this pH change are devastating to the animals and plants living in this ecosystem. These ecologically disasterous effects therefore result from the activities of countries over which the recipient of the pollution has no control and this has resulted in a great deal of argument between nations and vain attempts to formulate international law to reduce the problem in future years.

What is the effect of acidification? There are many examples of the deleterious effects of acidification upon animals: perhaps the best known relates to fish. Acid rain falling onto the land leaches metals from the soil and carries them to rivers and lakes. Aluminium is leached in this way and as a result of acid rain its concentration in the waterways in affected areas has increased considerably. The aluminium remains in solution in the acid water; however, when it comes into contact with fish gills, where the pH of the microenvironment is higher, aluminium salts precipitate and reduce the efficiency of gaseous exchange. Eventually the fish is unable to 'breathe' efficiently and dies.

There are many other effects of acid rain. Acidification of streams results in changes in the range of animals and plants that are able to inhabit the ecosystem. These changes do not form an endpoint in themselves, they might have a significant knock-on effect. For example, the dipper (*Cinclus cinclus*, a British bird which inhabits the banks of streams and brooks and feeds underwater on small animals (e.g. the common stonefly larvae (*Isoperla grammatica*)) which inhabit only neutral and slightly alkaline waters) is declining in certain areas because acidification of the waters has resulted in local extinction of the birds' preferred food.

The land is also affected. Trees (particularly conifers) have begun to die with a characteristic reduction in chlorophyll production. This has resulted in entire forests dying. Clearly the environmental impact of this is enormous because the trees are ecosystems in themselves, providing food and shelter for many animals and plants.

1.2.2   *Nitrogen Oxides and Nitrate*

Combustion of nitrogen-containing compounds (e.g. from oil) results in the release of a series of nitrogen oxides:

Nitrous oxide N_2O

Nitric oxide NO

Nitrogen dioxide NO_2

This class of air pollutant is referred to as NO_x or NOX for convenience and because in the environment there is the possibility of interchange between the individual oxides of nitrogen. They are all ultimately oxidised to NO_2 by the oxygen in the air:

$$N_2O + O_2 \rightarrow 2NO_2$$

$$NO + 1/2O_2 \rightarrow NO_2$$

NO_2 behaves rather like SO_2 because it dissolves in atmospheric water and is oxidised to nitric acid which adds to the acidification problem:

$$NO_2 + H_2O + 1/2O_2 \rightarrow HNO_3 + OH^-$$

Nitric acid or nitrate not only results in acid rain but also has a very different impact upon the environment. Nitrates are essential nutrients for plants and therefore increased concentrations, particularly in aquatic environments, can result in excessive plant growth which leads to vigorous plant growth out-competing other inhabitants of the ecosystem. This is termed eutrophication and is a serious problem in some agricultural areas where farmers are rather too enthusiastic with their use of nitrate fertilisers. On the face of it you might think that increased plant growth is good for an ecosystem because photosynthesis would result in the generation of oxygen which is important. The problem is that during the hours of darkness the plants respire so consuming valuable oxygen and deplete valuable nutrients (e.g. minerals) at all times. Sometimes the growth is so prolific that it is simply not possible for higher animals to live in the ecosystem (e.g. fish cannot swim). Eventually the plants exhaust their environment of essential nutrients and they die. Their decay increases biological oxygen demand (BOD) because the bacteria responsible require oxygen to live; this in turn reduces oxygen levels and so prevents fish and other oxygen-requiring species from inhabiting the ecosystem. It is very important indeed that eutrophication is controlled before it has irreversible effects on a significant number of aquatic environments.

The Norfolk Broads on the eastern coastal region of the UK and many rivers in Wisconsin in the USA provide good examples of eutrophication. Reed and other aquatic plant growth has become so great that silt has been trapped which in turn has reduced water flow and is slowly changing the ecosystems' characteristics. The reason for these parts of the world being particularly prone to eutrophication is that they are important farming areas (Wisconsin is America's Dairy State and Norfolk is an important arable region in the UK) and chemical fertilisers rich in nitrate and phosphate are used, or perhaps over used. There are, of course, other impacts upon this specific environment which are arguably more important. For example, it is a popular amenity area for holiday makers: their boats' propellers churn up the environment, disturbing sediments, killing plants, disrupting fragile banks and the exhausts expel toxic diesel and petrol into the water.

Nitrate from fertilisers is just one example of pollution, there are many more diverse examples. Agricultural waste (e.g. cattle and pig slurry) and human sewage are very important sources of nitrate.

1.2.2.1 Effects of NO_x on Humans

There is concern about the effects of airborne nitrogen oxides upon human health especially in cities. Studies have suggested that when NO_x concentrations in the atmosphere rise that there is a concomitant increase in asthma attacks. On the face of it this points to asthma being caused by NO_x; however, very recent studies suggest that NO_x actually sensitises the airways to the allergenic effects of, for example, pollen. NO_x is therefore a sensitisor not an allergen *per se*.

1.2.3 Sewage

In England in Tudor times (i.e. the period of the Tudor monarchy, 1405–1603) human excrement was thrown into the street where it was carried via open channels in the middle of the street (examples of these can be seen in Frome in Somerset in the UK) to rivers and streams where it was diluted and therefore had very little impact upon the environment. As the population rose the problem with human waste also rose until the health hazard posed by emptying the chamber pot out of an upstairs window directly into the street was realised. The realisation that disease was spread in this way led to the development of sewerage systems in large cities. The Victorians (i.e. people of the period of Queen Victoria's reign in Great Britain, 1837–1901) perfected these massive underground systems to transport waste directly into rivers and the sea. Later it was appreciated that there was a significant health hazard from the water in the rivers into which the sewage was poured. The need to treat the sewage became greater as systems were developed to extract water from rivers to supply whole cities with their drinking water. No one had yet realised that there might be ecological effects of the disposal of excrement into rivers or the sea or that the slurry derived from sewage treatment works might also pose an ecological problem. In the late twentieth century these ecological issues have become very real indeed and are the subject of heated debate at both local council and government level.

Surprisingly sewage is still pumped directly into the sea and results in levels of pollution on some beaches (e.g. those surrounding Morecambe Bay on the northwest coast of England) which exceed the newly introduced European Union

(EU) standards. These standards relate to bacterial pollution, but what about the chemical components of sewage? Under EU legislation it became illegal to dump sewage into the sea after 1997.

Before we consider the chemical composition of sewage it is important to understand sewage treatment processes.

1.2.3.1 Sewage Treatment

Household and industrial waste are taken via underground pipes to the sewage treatment works where they are allowed to stand in settling tanks to facilitate separation of heavy sediments. The supernatant is pumped into open systems (of several types, e.g. activated sludge) which allows bacteria to act upon the waste and degrade many of its organic components. The product of this degradative process is filtered through gravel beds and the treated water released into a river which eventually finds its way to the sea. The sludge which settles during and after the bacterial degradation phase of the process is termed sewage sludge; many millions of tons are produced annually in most developed countries in the world. The question is, what to do with this sewage sludge? Generally it is buried in or spread onto agricultural land as a fertiliser. There are even moves afoot to produce potting composts based on human sewage sludge – the salesmen of such products might, however, have quite a problem persuading potential customers to use them!

1.2.3.2 Sewage Sludge as a Pollutant

Many of the components of human excrement are concentrated in sewage sludge. For example, it is rich in nitrates (which is one of the reasons that it is a useful fertiliser). The high concentrations of nitrate pose extra problems of eutrophication (see above).

Humans (and other mammals) eliminate heavy metals from the body via the bile which is secreted into the duodenum (see Figure 1.7). The metals are incorporated into the faeces and therefore find their way into the sewage sludge. Metals such as copper, lead, cadmium and mercury are found at relatively high levels in sludge. If sludge is used as a fertiliser the land becomes contaminated with these metals and the plants grown on the land might take up the metals and result in the consumers of the vegetables getting an unacceptably high dose of a particular heavy metal. The consumers then eliminate the metal in their faeces and the cycle begins again.

1.2.4 Agricultural Waste

As the human population increases so its need for food increases. As a major component of our diet is meat the need for farm animals correspondingly increases.

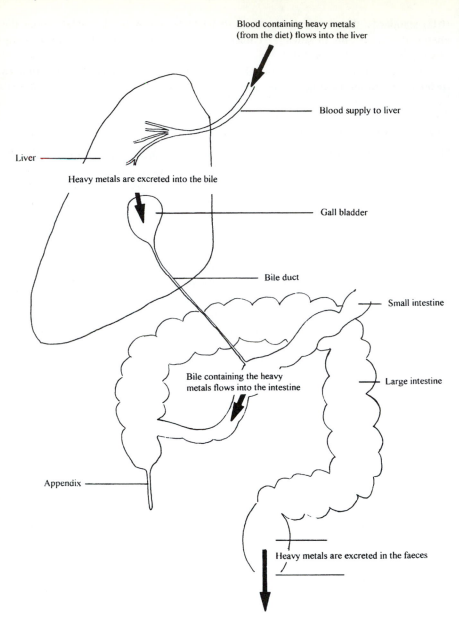

Blood containing heavy metals
(from the diet) flows into the liver

Blood supply to liver

Liver

Heavy metals are excreted into the bile

Gall bladder

Bile duct

Small intestine

Bile containing the heavy
metals flows into the intestine

Large intestine

Appendix

Heavy metals are excreted in the faeces

Figure 1.7 Schematic representation of the excretion of heavy metals in faeces via the bile.

More farm animals means more manure. The problems discussed above in relation to human excrement also apply here, but the solutions are very different.

There is no sewerage system for farm animal waste; instead the animals are either allowed to deposit it directly onto the land or (for intensively farmed animals, e.g. pigs) the slurry is collected in tanks or lagoons, allowed to stand for a while (a week or so) and then pumped onto the land. The problems associated with sewage sludge are similar to those of agricultural slurry. Slurry is rich in nitrate and contains relatively high levels of heavy metals. There is, however, an additional problem when slurry is applied to agricultural land and that is that if the animals from which the slurry has been derived have been dosed with veterinary medicines (either as treatments for disease or as growth promoters) residues of the drugs and their metabolites will be present in the slurry. The land then becomes contaminated with the drugs and crops grown on it might contain low levels of the drugs. This presents a very low risk indeed to the consumer because the residues in food plants are extremely low; however, ecologically there is potentially more of a problem. Let us consider a hypothetical example. Slurry derived from a pig farm where the pigs have been treated with sulphadimidine (a sulphonamide) to eradicate a respiratory infection on the farm is applied to the land. Sulphadimidine is an antimicrobial agent and so will modify the natural bacterial flora of the soil; over the years this might result in a quite significant and permanent change in soil organisms. The soil organisms are very important indeed in supporting the terrestrial ecosystem; they degrade organic matter, fix nitrogen, etc. Unnatural changes are undesirable.

There is a very good example of the effects of interference with the natural decay process caused by veterinary medicines in animal faeces. This example relates mainly to faeces deposited directly onto the field rather than collected as slurry. In the summer months when cattle are grazing they produce vast

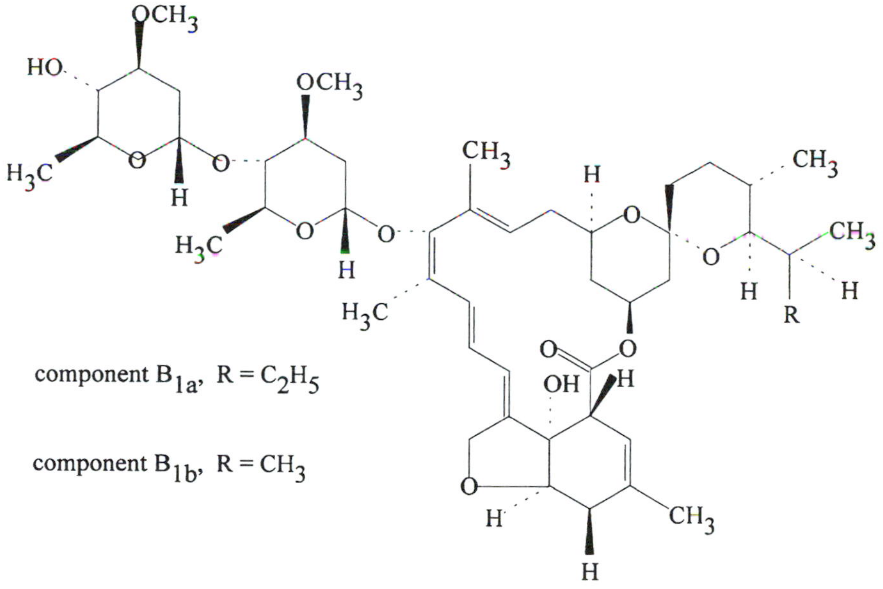

component B$_{1a}$, R = C$_2$H$_5$

component B$_{1b}$, R = CH$_3$

Figure 1.8 Molecular structure of the nematocide, Ivermectin.

15

quantities of manure; despite this the fields do not disappear under this excrement. The reason for this is the decay and the removal of the manure by worms, nematodes and dung beetles (among many other species). It takes about a week for a cow pat to disappear from a grassy field. Recently, however, some cow pats have persisted for very much longer due to the use of the nematocide Ivermectin (see Figure 1.8) to treat cattle for nematode infestation. Ivermectin kills the worms, nematodes and dung beetles so effectively that degradation of the cow pat is inhibited.

1.2.5 *Warfare*

No discussion of the environmental impact of human activities would be complete without a consideration of the devastating effects of war. Some of these effects are obvious; enormous areas of land are laid waste by bombs. Unthinkable atmospheric pollution results from, for example, the massive oil fires that resulted from the Gulf War in 1990. Some jingoistic activities, however, result in rather more esoteric ecological effects.

During the Vietnam War in the 1970s the American forces had difficulties operating in the dense jungles. Their enemy was well aware of this and took (very sensibly) advantage of the situation. The Americans overcame the problem by the use of a herbicide called Agent Orange. Agent Orange is the trivial name for 2,4,5-trichlorophenoxyacetic acid (2,4,5-T; see Figure 1.9) which is an auxin agonist that speeds up the growth of plants resulting in premature senescence and leaf drop.

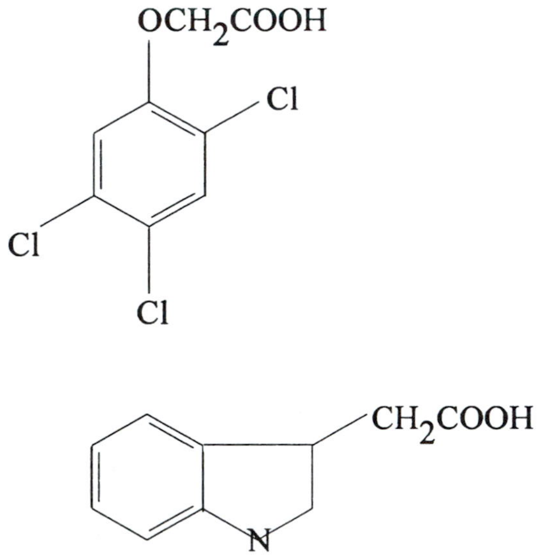

Figure 1.9 Molecular structure of 2,4,5-T (top) showing its structural analogy with the plant growth hormone, auxin.

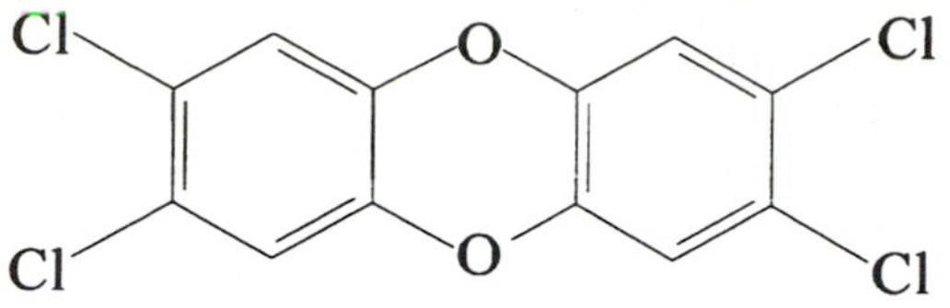

Figure 1.10 Molecular structure of 2,3,7,8-tetrachlorodibenzo[b,e][1,4]dioxin (TCDD), the dioxin which contaminated Agent Orange. Its LD_{50} (rat, male) = 0.022 mg kg^{-1}.

Once the leaves had fallen off the forest trees the Americans could see their enemy from the air.

The environmental effects of this use of 2,4,5-T were devastating because of the effects upon the forest plants. The forest is a very carefully balanced ecosystem which easily becomes unbalanced when one of its component parts is removed. For example, plants growing on the forest floor are suppressed by the reduced light level caused by the tree canopy; however, when defoliation resulted from 2,4,5-T use the forest floor flora changed. This had a knock-on effect resulting in a change in the fauna, some of which depended on specific plants growing on the shady forest floor. In fact as a result of the Vietnam War 60% of the defoliated forest has now been replaced by scrub land which represents an early step in recolonising the land. It will take hundreds of years for the diverse forest environment to re-establish itself.

There was a rather more sinister problem associated with Agent Orange. The enormous amounts needed for the Vietnam War resulted in a rather less chemically pure preparation being manufactured. This impure Agent Orange contained traces of highly toxic dioxins (see Figure 1.10) which are thought to be carcinogenic to humans. Indeed, there have been many accusations of ill health in both American soldiers and Vietnamese victims following their exposure to dioxin-containing Agent Orange.

War is generally viewed in a very negative sense; however, there are some positive environmental outcomes of such devastation. Areas dotted with unexploded mines become havens for wildlife because humans dare not enter. Bomb craters often fill with water and turn into diverse aquatic-based ecosystems. Every cloud has a silver lining!

1.2.6 Pesticides

1.2.6.1 Organophosphorus Pesticides

The first organophosphorus pesticides (OPs) were synthesised by Lassaigne in 1820; however, the true precursors of present day OP insecticides were not manufactured until 1854 when Clermont first synthesised tetraethyl pyrophosphate (TEPP; see Figure 1.11). Clermont did not realise the powerful insecticidal properties of TEPP, indeed 80 years passed before the extreme toxicity of OPs was investigated. World War II provided the impetus to investigate the possibility that

17

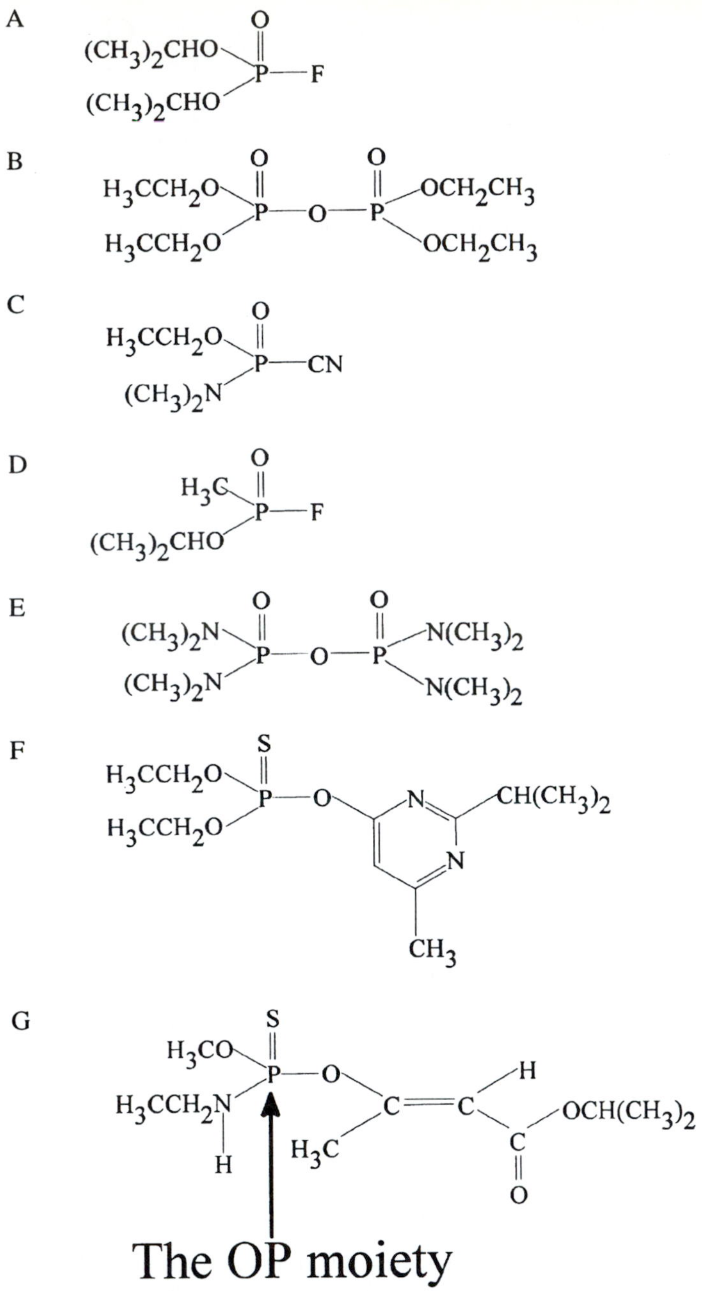

A

$(CH_3)_2CHO$ — P(=O) — F, $(CH_3)_2CHO$

B

H_3CCH_2O — P(=O) — O — P(=O) — OCH_2CH_3, H_3CCH_2O, OCH_2CH_3

C

H_3CCH_2O — P(=O) — CN, $(CH_3)_2N$

D

H_3C — P(=O) — F, $(CH_3)_2CHO$

E

$(CH_3)_2N$ — P(=O) — O — P(=O) — $N(CH_3)_2$, $(CH_3)_2N$, $N(CH_3)_2$

F

H_3CCH_2O — P(=S) — O — [pyrimidine ring with $CH(CH_3)_2$, N, N, CH_3], H_3CCH_2O

G

H_3CO — P(=S) — O — C(CH_3)=C(H) — C(=O) — $OCH(CH_3)_2$, $H_3CCH_2N(H)$

The OP moiety

Figure 1.11 Molecular structures of diisopropylfluorophosphate (A), tetraethylpyrophosphate (B), tabun (C), sarin (D), octamethylpyrophosphoramide (E) and the sheep dip pesticides Diazinon (F) and Propetamphos (G). The basic OP structure can be clearly seen to be common to them all.

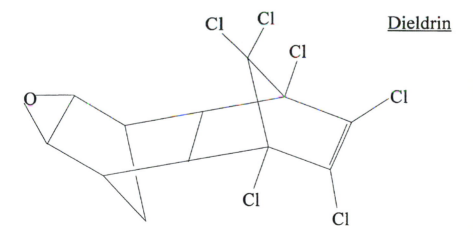

DDT

CCl₃
|
Cl—⟨ ⟩—CH—⟨ ⟩—Cl

Aldrin

Cl   Cl
Cl

Cl

Cl

Cl

Dieldrin

Cl   Cl
Cl

Cl

O

Cl

Cl

Structure of Dieldrin

Figure 1.12   Molecular structure of DDT and the related pesticides, aldrin and dieldrin, showing their hydrophobic nature.

OPs might be used as nerve gases in chemical warfare; Saunders and his colleagues studied alkyl fluorophosphates (see Figure 1.11) as potential nerve gases (interestingly at the same time German scientists were studying tabun (see Figure 1.11) and sarin (see Figure 1.11) with the same purpose in mind).

The extreme mammalian toxicity of OPs precluded their use as insecticides. In 1941, however, Schrader prepared octamethylpyrophosphoramide (see Figure 1.11) and demonstrated its insecticidal properties. From this beginning has developed an enormous range of OPs which are used as important insecticides in agriculture and as ectoparasiticides in veterinary medicine.

1.2.6.2 *Dichlorodiphenyltrichloroethane*

Dichlorodiphenyltrichloroethane (DDT; see Figure 1.12) was among the first effective pesticides to be synthesised. It was first synthesised in 1874, but its insecticidal properties were not discovered until 1939 by a Swiss scientist. It was not used widely until immediately after World War II and soon became established because of its amazing effectiveness. At the time there were no other pesticides on the market with such a high degree of efficacy (and apparently low toxicity – DDT has a low acute human toxicity and when compared with the intensely toxic alternatives (e.g. nicotine) of the time it appeared to be very safe). The problem was that no one realised the potential environmental toxicity of DDT; for every benefit there is a drawback and DDT was a severe lesson to be learnt. This, of course, is now all history.

Further Reading

Clapp, B.W., 1993, *An Environmental History of Britain*, Harlow: Longman.
Moriarty, F., 1983, *Ecotoxicology*, 2nd Edn., London: Academic Press.

2

Effects of Pollutants on Ecosystems

This chapter introduces general principles of ecology, including food chains, food webs and energy flow in ecosystems.

The impact of synthetic chemicals upon the environment is covered with reference to pesticides:

- DDT and eggshell thinning
- Dieldrin and herons
- Tributyltin oxide and imposex in dog whelks
- OCs and bats

and other pollutants

- Nitrates and algal blooms
- Clenbuterol residues in meat
- PCBs and immunosuppression

The need to assess the risk and benefit of chemicals and how this has been used to stimulate the development of safer pesticides is introduced.

2.1 An Introduction to Food Webs

Before it is possible to understand the effects of pollutants upon ecosystems it is essential to have an appreciation of food chains and food webs. This section is not intended to be a comprehensive discussion of food chains and webs, but simply to outline their organisation and the routes by which foreign chemicals find their way into animals and plants and more importantly concentrate up the food chain. Jonathan Swift, the eighteenth century satirist, summed up food chains admirably in this verse:

So, naturalists observe, a flea
Hath smaller fleas that on him prey;

21

And these have smaller fleas to bite 'em
And so proceed ad infinitum.
Thus every poet, in his kind
Is bit by him that comes behind.

Humans, unlike most other animals and plants, dominate and control the environment in which they live and thus seemingly control the balance of all nature. The human population is growing at an alarming rate, apparently unchecked. Human ingenuity in the field of science is now capable of manipulating the basic genetics of life in order to change organisms to better serve our needs. Human population growth is not infinite and cannot go unchecked; many have said that genetic interference in the processes of nature may have dire consequences. On closer examination we find that people have existed for only a brief time in the evolution of life on this planet, appearing no more than a million years ago. Indeed, all warm blooded animals and other vertebrates have existed for no more than 450 million years in this planet's four billion year history (i.e. 0.01% of the time since life began). Simple living organisms developed in the earth's primeval seas about three billion years before the appearance of people. Fossil records do not exist before this time.

Early life was probably similar in function, shape and form to the bacteria and viruses of today, deriving its energy by fermenting organic molecules that surrounded it in abundance in the primeval oceans. The oldest fossil discovered, *Eobacterium isolatum*, was found in South Africa in sedimentary rock and is over three billion years old. *E. isolatum* was a prokaryote (pre-nuclear) just like all bacteria living today. In other words it was a simple cell without organelles (often classified in a separate kingdom called Monera). Indeed, the metabolic ability of many present day bacteria would have suited them well for survival in the primitive oceans. Many species live on sediment of oceans and manufacture organic compounds from CO_2 and H_2S using the energy from sunlight.

$$CO_2 + 2H_2S \xrightarrow{\text{light}} (CH_2O) + H_2O + 2S$$

It is interesting that recently (late 1995) new species of deep sea sulphur bacteria have been discovered which live at high temperatures and pressures around volcanic vents. These are probably very similar to those of the primordial oceans – in fact they might be the very organisms that were present in prehistoric waters which have clung on to this very specialised and hostile deep sea environment.

The fact that primitive organisms have survived unchanged reflects that conditions for their survival have remained stable and optimum with little evolutionary pressure. Life did evolve, however, and from those early primitive bacteria more complex organisms developed and the form and function of life became more diverse. First came simple plant and animal cells. These were the first eukaryotes (having a nucleus). Simple single celled plants and animals evolved into multicellular organisms which began the complex process of evolution that resulted in everything we see today.

As animals and plants developed they coexisted with the more primitive forms and indeed they became dependent upon each other. Diversity increased and complicated feeding associations developed. These associations are referred to as food chains and food webs. The driving force within these food chains and webs is one of survival and adaptations to the pressures of life. Jonathan Swift recognised this relationship, however naively, before Charles Darwin wrote *Origin of the Species by Means of Natural Selection* in 1859 which was more appropriately entitled the *Preservation of Favoured Races in the Struggle for Life*. He discussed in depth the effects of competition and recognised the complex interrelationships between animals and plants, and concluded that *the relationship of organism to organism is the most important of all relations.*

Modern day thinking on food webs fits in well with Darwinian theory. In simple terms a food web explains the transfer of energy from primary producers (organisms which derive energy directly from sunlight, i.e. plants) to complex consumers (carnivores who eat carnivores, e.g. eagles). The stages or levels of feeding are referred to as trophic levels (trophic comes from the Greek *trophicos* meaning nourish) (see Figure 2.1).

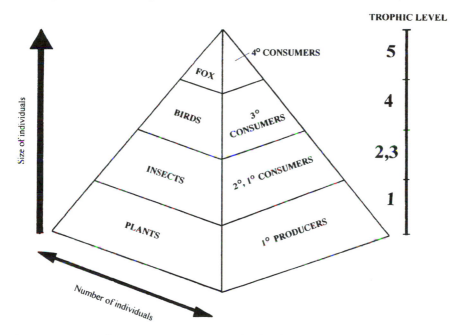

Figure 2.1 Pyramid representation of the trophic levels and their interactions.

2.1.1 *The Organisation of Food Webs*

Food chains are dependent upon primary producers which input energy. This energy is transferred up the food chain. This primary energy input is derived from photosynthesis in which CO_2 is converted to complex carbohydrates utilising the sun's energy. It has been estimated that plants fix about 35×10^{15} kg of carbon per year:

$$\text{sunlight}$$
$$CO_2 + H_2O \longrightarrow CH_2O + O_2$$

The primary producer (e.g. grass) is eaten by primary consumers (e.g. cows) and the energy originally derived from the sun is transferred to the next trophic level. The problem is that energy transfer is not very efficient, a large proportion of the energy assimilated is lost as heat and therefore only a relatively small proportion is transferred up the food chain.

The concept of the food chain is quite simple, but alas in reality it is rather more complex. For instance, humans could be primary consumers if they were vegetarians, on the other hand if they included beef (or any other herbivore) in their diet they would be secondary consumers and if they liked mackerel they would be tertiary consumers! It is conventional to classify a species with respect to its highest trophic level. People, however, do not fit well into this simple approach because they are opportunists and will eat what they need to survive or what suits their religious beliefs. In general people are regarded as tertiary consumers.

As discussed, the fundamental concept of the food web relates to energy transfer and therefore food webs are governed by the laws of thermodynamics.

2.1.1.1   The First Law of Thermodynamics

Energy can be transferred from one type to another but cannot be created or destroyed.

Sunlight is a form of energy and can be transformed into work, heat or the potential energy of food. A grazing animal will consume the carbon components and sugars stored in the grass; its enzymes will break down these compounds creating the energy required to maintain its living processes by respiration.

$$C_6H_{12}O_6 + 6O_2 \rightarrow 6H_2O + 6CO_2 + \textbf{ENERGY}$$

In the process some of the energy is lost as heat.

2.1.1.2   The Second Law of Thermodynamics

No process involving an energy transformation will spontaneously occur unless there is a degradation of the energy from a concentrated form into a dispersed form.

As discussed above, the loss of energy as heat is an important factor in the inefficiency of energy transfer up the food chain. This inefficiency of energy transfer is why food chains are represented as pyramids (see Figure 2.1) because an enormous biomass of organisms is necessary at the bottom of the food chain to support a small biomass at the top.

The concept of trophic levels has been in existence since 1927 when Charles Elton referred to his pyramid of numbers. In his youth Elton went on an expedition to Bear Island in the Arctic Circle where he studied the ecology of the tundra. The harsh conditions provided the ideal setting to study simple food chains and animal and plant feeding relationships. He studied the community in which the Arctic fox was top of the food chain in an environment where there was no tall vegetation. There were few animals and plants and all the feeding activities were easily observable. He noted that foxes caught the summer birds of the tundra, for example sandpiper, ptarmigan and bunting. The ptarmigan ate berries, leaves and tundra vegetation; this food chain consisted of tundra-ptarmigan-fox. In the same environment sandpipers ate insects which in turn ate tundra plants; this is a separate food chain. In addition, foxes also ate seabirds such as gulls and eider ducks, hence the fox was also dependent upon the sea and sea plants. In the winter when birds were not available, foxes fed on polar bear dung and the remains of seals killed by polar bears. Polar bears and seals were part of the aquatic food chain. It is clear from this example of a feeding interrelationship that simple food chains do not account for the complexities of nature. In fact separate food chains are linked to form food webs (see Figure 2.2).

In this example, the fox was the largest land animal followed by the birds which the fox caught. The birds were both smaller and more numerous. The birds also fed on insects which were very much smaller and very much more numerous. On working up the food web it became apparent to Elton that its members became progressively fewer but larger. Of course, there is a maximum size for an animal on purely physical grounds. For example, an animal the size of the blue whale lives in the sea because it would probably be too heavy to move around on land. Another

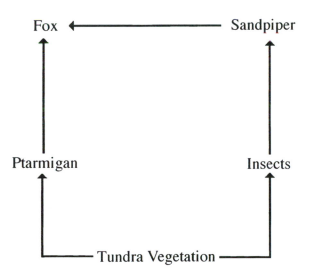

Figure 2.2 Schematic representation of the interrelationship between two food chains to form a simple food web as first described by Elton in 1927.

TROPHIC LEVEL

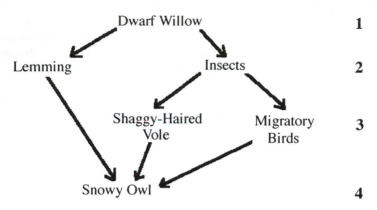

Figure 2.3 Complex food web showing the trophic levels.

limitation on size is the need to find sufficient food to support a large mass; the larger an animal is the further it has to roam to catch enough food to survive, clearly there is a limitation here because a large amount of energy is used roaming and there comes a point at which more energy is used finding food than is obtained from it!

To add to the complexities of the food web it is very likely that individuals will eat more than one food (e.g. foxes eat rabbits and birds) and therefore the complexity of the food web must be increased to account for this (see Figure 2.3).

A complex food web is composed of only two basic food chain types, namely the grazing food chain and the detritus food chain. Furthermore, all organisms can be classified into one of only seven trophic levels (see Table 2.1).

Table 2.1 The seven trophic levels showing examples of organisms

Trophic level	Description	Example
1 (P)	Primary producer	Spyrogyra, oak tree
2 (C1)	Primary consumer (herbivores)	*Daphnia*, elephant
3 (C2)	Secondary consumer (carnivores)	Water spider
4 (C3)	Tertiary consumer	Trout, wolf
5 (C5)	Quarternary consumer	Birds of prey
S	Saprophytes	Bacteria, fungi
D	Decomposers	Bacteria, earthworm

It is important to remember that the diversity within a trophic level is very great. For example, both elephants and *Daphnia* are in tropic level 2.

2.1.2 The Grazing Food Chain

This involves a flow of energy directly from living organisms. A typical example of a grazing food chain is an African plain ecosystem where energy input is from sunlight; this is assimilated by grass to make nutrients which are consumed by grazing herbivores (e.g. antelope), and carnivores (e.g. lions) lie in wait to eat the herbivores (see Figure 2.4).

The members of the grazing food chain produce waste and eventually die and so contribute to the detritus both during their life and at its end. There is a separate organisational food chain (the detritus food chain) which clears up this waste and returns the nutrients to the ecosystems in accessible forms.

GREEN PLANTS ➔ GRAZING HERBIVORES ➔ CARNIVORES
(predators)

e.g. Grass   e.g. Antelope   e.g. Lion

Figure 2.4 Grazing food chain showing how energy flows from primary producers via primary consumers to the predatory carnivores (in this case secondary consumers).

2.1.3 The Detritus Food Chain

Here the energy from decaying animals and plants or the excrement from animals flows back into the food chain with the help of microorganisms (see Figure 2.5). A good example of a detritus food chain is a compost heap where a rich bacterial flora degrades the organic matter (e.g. grass clippings); unfortunately energy is wasted as heat here (put your hand onto a compost heap and it will feel very warm). When you dig into the compost you will find many earthworms and when you use the compost a robin or blackbird will not be far away hoping for an earthworm meal.

In reality these two types of food chain are rarely spatially or temporally separated. In any food web there will always be some photosynthetic or chemosynthetic production together with natural cell death and decay. As a general rule energy flow in an aquatic environment is greater via the grazing than by the detritus pathway. Death within a food web will always return some energy to the trophic pyramid.

2.1.4 The Importance of the Interrelationship Between Species in Environmental Toxicology

A temperate woodland in summer might contain only 200 mature trees, but they might support a population of 15 000 000 primary consumers such as leaf-eating invertebrates. The reason for this is not difficult to understand as each mature tree will contain a vast biomass in terms of consumable leaf area in summer when large

27

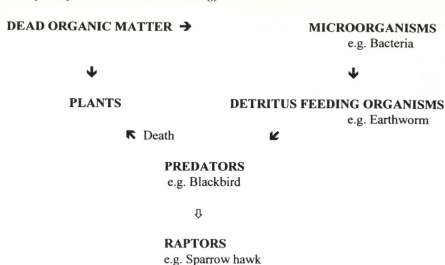

Figure 2.5 Detritus food chain showing how dead organic matter is returned to useful service in an ecosystem. In this case the energy from the dead organic matter enters the earthworm via microorganisms. The earthworms are eaten by blackbirds which themselves might be preyed upon by sparrow hawks. The whole process begins again when the sparrow hawk dies.

transient populations can feed. In turn, insectivorous animals and birds will feed on them. In autumn and winter dead and decaying leaf litter will provide food for detritivores which will provide a food source to maintain populations at higher trophic levels. Thus the biomass pyramid gives a better picture of food web relationships for an ecological group as a whole than a food web. If a toxic chemical is introduced at the bottom of the biomass pyramid there is likely to be a significant concentration effect as one moves up the trophic levels. It is therefore inevitable that the tertiary consumers will succumb to the toxic effects of the chemical. The classic example of this is DDT and its effects upon birds of prey. A knowledge of feeding mechanisms can often enable us to predict the extent of a potential toxic effect of a chemical in the environment (i.e. environmental impact). The role of food chain concentration in the toxicity of DDT to birds of prey is illustrated in Table 2.2.

2.2   Some Specific Examples of the Environmental Toxicities

2.2.1   *Dichlorodiphenyltrichloroethane (DDT) and its Environmental Impact*

Dichlorodiphenyltrichloroethane (DDT; see Figure 1.12) is a very hydrophobic molecule which acts by interfering with ion transport systems in the neuronal cell membrane. This interference inhibits neurotransmission and so kills animals if

Table 2.2   Food chain concentration of DDT in an east coast estuary in the USA

	Diet	DDT residues (parts per million)*
Water	N/A	0.00005
Plankton	N/A	0.04
Sheepshead minnow	Plankton	0.94
Pickeral	Predatory fish	1.33
Heron	Small fish	3.57
Herring gull	Scavenger	6.00
Osprey (eggs)	Larger fish	13.8
Merganser	Fish	22.8
Cormorant	Larger fish	26.4

Data from Woodwell, G.M., Worster, C.F.J. and Isaacson, P.A. (1967), *Science*, **156**, 821.
* Parts per million $= $ mg.kg^{-1} or mg.dm^{-3}.

given at a sufficiently high dose. As all animals rely upon ion exchange across the neuronal membrane to initiate an action potential (and therefore neuro-transmission), DDT is not species specific in its effects.

DDT and the related insecticides, endrin, dieldrin and aldrin, are termed organochlorine pesticides. They are all (see Figure 1.12) heavily chlorinated hydrophobic ion channel inhibitors and are very toxic indeed.

DDT was introduced in the 1950s and was hailed as a miracle; it was the first real pesticide and it revolutionised farming practice throughout the world. Crops could be grown without the problems associated with insect damage. And what is more DDT was safe! There is a famous section in a television documentary film in which a man eats DDT powder from his hands and apparently suffered no ill effects. This was in the halcyon days when little attention was paid to the concept of chronic toxicity. I wonder how the DDT eater is 40 years later. DDT was very commonly used indeed with some 4×10^5 tons being manufactured worldwide in 1964; its use declined as we became aware of its environmental toxicity (see Table 2.3).

The case of DDT is a classic example of our ignorance in failing to predict the effects of the widespread use of this highly persistent pesticide on the environment. DDT was the first of a new age of synthetic organic chemicals developed after World War II. Many other chlorinated hydrocarbons followed. It is probably the most notorious pesticide that has ever been on the market.

The story of its rise to stardom, carrying with it the Nobel Prize, and decline to infamy is quite sensational ...

George W. Ware, 1974.

First synthesised over 120 years ago by a German graduate student, the formula lay forgotten for almost 70 years until rediscovered by a Swiss entomologist, Dr Paul Müller when searching for an insecticide for use against the common clothes moth. The effectiveness of DDT exceeded his wildest dreams; not only did it kill the clothes moth, it also proved effective against many other insects, particularly those

Table 2.3 Global manufacture of DDT since 1964, showing its decline following the realisation that it was extremely toxic to the environment after prolonged use

Year	Global production of DDT ($\times 10^5$ tons)
1964	4
1967	3.3[a]
1971	2–2.5

[a] This figure includes DDT plus some other OCs.
Data from Brooks, G.T., 1974, *Chlorinated Insecticides*, Vol. 1, Ohio: CRC Press.

mosquito species that carry malaria and yellow fever, body lice that transmit typhus fever and fleas which are vectors of the plague. For this discovery he received the Nobel Prize in 1948. From then to the mid-1960s it was seen as the 'magic bullet' for a whole range of insect pests, not only vectors of human disease but also of agricultural pests. Of the 1.8×10^9 kg per year of DDT used worldwide 80% was used in agriculture. It has been used against the Colorado potato beetle, the apple codling moth, cotton bollworm, the gypsy moth and the spruce budworm. It was extremely cheap and was without doubt the most economical pesticide ever sold. Despite all this it was banned in the USA in January 1973 and the rest of the world soon followed; so what went wrong?

As is so often the case DDT was a victim of its own success. It was so effective and so cheap that nobody wanted to use anything else. If we had studied the compound a little more carefully with a little knowledge of the relationships between organisms in a food web then the catastrophic effects seen in some species could have been avoided, the pest species it was supposed to control would not have become resistant and the compound could have still been in use today.

Its mode of action is less important to this discussion, in any event it is not fully understood. It is thought to inhibit the Na^+-K^+ pump, which is crucial to neurotransmission, causing spontaneous firing of neurons associated with involuntary muscle twitching which eventually leads to death. The most important fact is that the molecule is extremely stable and is therefore extremely persistent in water, soil, animal and plant tissue. The half-life in soil is in the region of 30 years. Its major metabolite, dichlorophenyldichloroethane (DDE), is even more persistent in animal and plant tissue and the environment. DDT and DDE are only extremely slowly degraded by microorganisms, heat or ultraviolet light. DDT has a very low water solubility ($6\,ng\,dm^{-3}$). It is therefore extremely lipophilic and preferentially accumulates in the fat components (e.g. phospholipid cell membranes) in plant and animal tissues. As it is neither metabolised nor excreted and is freely stored in body fat it accumulates at each level in the trophic pyramid.

The first evidence that all was not well with DDT came, surprisingly, from the pigeon racing fraternity in 1960. Pigeon fanciers maintained that their prize birds were being killed more often by peregrine falcons than in the past. They concluded that the falcon population was increasing. After investigation it transpired that the opposite was the case. A survey conducted to resolve the issue showed there had been a dramatic reduction in falcon numbers in southern England which then spread northward. This catastrophic decrease in population resulted in a 92% reduction in its territorial range.

Raptor (predatory birds such as eagles and peregrine falcons) populations began to decline too. Extensive epidemiological studies demonstrated that this decline in numbers was associated with eggshell thinning. The thin shells meant that eggs were likely to be broken in the nest and therefore not bring forth offspring. This phenomenon was potentially devastating to these already threatened (because their preferred environments were also declining) species. An association between ingestion of DDT and eggshell thinning was demonstrated as was the fact that raptors were at the top of the food chain and would accumulate a large body burden of hydrophobic pesticides such as DDT. Clearly the raptors were at greatest risk from the chronic toxicological effects of DDT.

The mechanism of DDT-induced eggshell thinning is still not fully understood; however, it is thought to involve the inhibition of carbonic anhydrase by DDT (and perhaps DDT metabolites). Carbonic anhydrase catalyses the formation of CO_3^{2-} from CO_2, particularly in the egg formation process because eggs are composed chiefly of $CaCO_3$. DDT-thinned (achieved by adding $100\,mg\,kg^{-1}$ of DDT to the diet of experimental quail) egg shells contain significantly less $CaCO_3$, which supports the theory.

DDT was effectively banned in the mid-1960s and is now very rarely used except in very special cases where its benefits far outweigh its risk to the environment. One such use is the eradication of the malaria-carrying *Anopheles* mosquito in Africa. Without DDT it would be impossible to control malaria and many people would die. Its risk to the environment when used in these very small quantities is irrelevant in the context of the human benefit.

Very recently problems associated with DDT have resurfaced. Rabbit meat imported from China has been found (by the UK Government's Working Party on Pesticide Residues (WPPR), a working party of the Advisory Committee on Pesticides (ACP) and the Food Advisory Committee (FAC) which monitors pesticides in food and advises ministers on their findings) to contain residues of DDT which indicates its use in China. This illustrates the enormous difficulty of controlling the use of pesticides worldwide. The UK can stop the import of Chinese rabbit meat in order to protect our consumers, but what about the global environment? DDT is very mobile and is now beginning to turn up in environments far from China. One country's unacceptable risk is of no concern to others.

2.2.2 *Dieldrin and Herons*

Like other OCs dieldrin has proved toxic to wildlife. After routine monitoring of wildlife by the UK Government in the 1980s high concentrations of dieldrin were found in eels caught in the rivers of southern England. Independently a large number of herons were being reported dead in the area. On post-mortem examination it was found that the herons also had high levels of dieldrin in their tissues. The cause of death was not determined, but OCs were thought to be, at least in part, responsible. This is a common problem worldwide because eels are very long lived (about 50 years) and therefore during this long life are able to accumulate large amounts of persistent pesticides.

As a result of the UK incident discussed above, a major company was heavily fined for releasing dieldrin into the environment. Subsequently, because of its toxicological profile, dieldrin was withdrawn by the UK and other countries.

2.2.3 *Organophosphorus Pesticides*

The mechanism of action of OPs relies upon their inhibition of the synaptic enzyme acetylcholinesterase (AChE; see Figure 2.6). This is to some extent insect specific because most insects rely upon cholinergic transmission, whereas other animals have alternative mediators of neurotransmission and therefore inhibition of the cholinergic pathway is not quite so devastating.

In 1938 the Sheep Scab Order was introduced in the UK in an attempt to control the debilitating and fatal effects of the highly infectious sheep scab mite. The Order stipulated that sheep must be dipped (see Figure 2.7) in an approved proprietory dip which was based on either the OPs Diazinon or Propetamphos (see Figure 2.7) or a selected pyrethroid insecticide (in the days before organic pesticide arsenicals and mercury compounds were used).

The Order remained in place for some 54 years until it was revoked in 1992. During the operation of the Order farmers became worried that they were suffering from the toxic effects of Diazinon or Propetamphos and consumers of lamb were concerned about the residues of the OPs in their diet.

The OPs have significant effects upon the environment; when used in sheep dips many thousands of litres of spent dip have to be discarded. Disposal is generally by burying or, more commonly, by spraying the dip onto fields. The rationale behind this approach is based upon the rapid degradation of the OPs by soil bacteria and when exposed to UV light. The degradation does, however, take time; the half-life ($t_{1/2}$) in soil is of the order of 5–10 days (depending on soil type, temperature, rainfall, etc.). During this time a great deal can happen. Perhaps the most significant hazard (or at least noticeable) is the toxicity of OPs to geese. Geese are gregarious grazing birds and there have been several examples of large numbers of geese dying as a result of feeding on a field upon which OP sheep dip has been discarded. It is uncertain why geese are particularly prone to the toxicity of OPs, but it is possible that they have a reduced metabolic capacity for OPs which means that they are detoxified more slowly and consequently are more toxic.

Concerns about the environmental impact of OP sheep dips have led to one manufacturer devising a system to destroy the OP before the dip is discarded. This simple system (devised by Grampian Pharmaceuticals Ltd, UK) relies upon

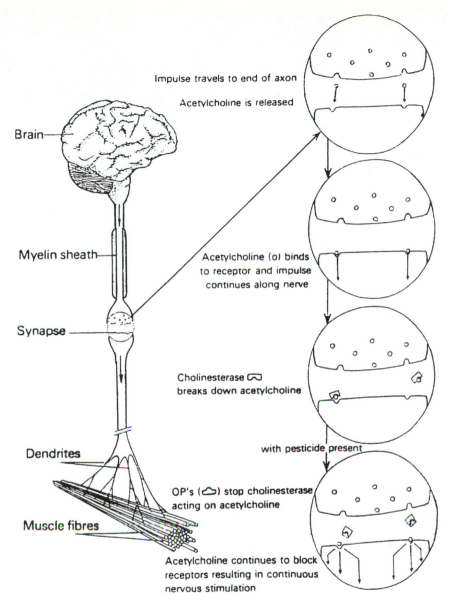

Impulse travels to end of axon

Acetylcholine is released

Brain

Myelin sheath

Acetylcholine (o) binds
to receptor and impulse
continues along nerve

Synapse

Cholinesterase ⌒⌐
breaks down acetylcholine

with pesticide present

Dendrites

OP's (⌒) stop cholinesterase
acting on acetylcholine

Muscle fibres

Acetylcholine continues to block
receptors resulting in continuous
nervous stimulation

Figure 2.6 Mechanism of cholinergic neurotransmission showing the toxicity of OPs which inhibit acetylcholinesterase (AChE). Drawn by Jason Weeks, Central Veterinary Laboratory, Weybridge, UK.

hypochlorite (bleach)-catalysed oxidation of the OP. One can only hope that this clever approach catches on.

OPs have many more applications; indeed sheep dipping, although contro-versial, is probably among their more minor uses. There is a vast array of OPs

Figure 2.7 Sheep being dipped in an OP bath to prevent sheep scab and fly strike. Inset (p. 35, top) are the molecular structures of the two OPs used in sheep dips, namely Diazinon and Propetamphos. Photograph kindly provided by Grampian Pharmaceuticals Ltd, Leyland, Preston, UK.

Diazinon

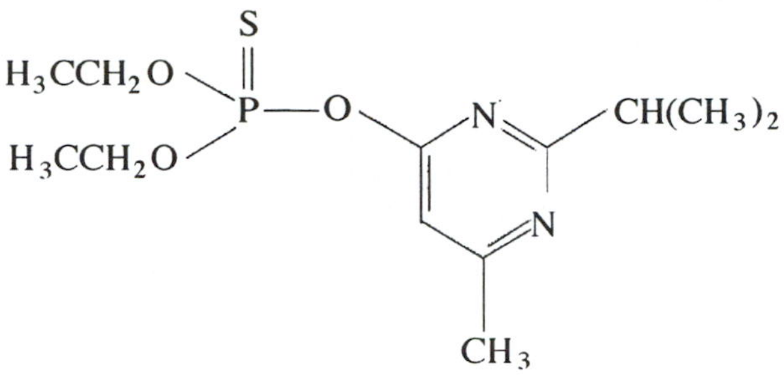

$$H_3CCH_2O - \overset{\overset{S}{\|}}{P} - O - \quad N \overset{}{=}\quad CH(CH_3)_2$$

(Molecular structure of Diazinon with P bonded to two H_3CCH_2O groups, double-bonded S, and O linked to a pyrimidine ring bearing $CH(CH_3)_2$ and CH_3 substituents)

Propetamphos

(Molecular structure of Propetamphos: H_3CO and $H_3CCH_2N(H)$ groups bonded to P with double-bonded S, P—O linked to $C(H_3C)$—$C(H)$ chain with $OCH(CH_3)_2$ and a $C=O$ group)

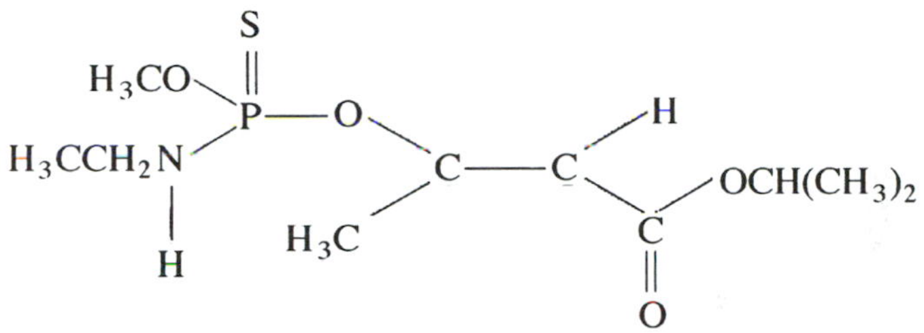

Molecular structures of Diazinon and Propetamphos.

designed for use as insecticides. Many thousands of tons are applied to crops each year to kill insect pests (e.g. aphids). The ecological problem is that these OPs are not specific for the pests, but kill beneficial insects too.

2.2.4   *Tributyltin Oxide and the Dog Whelk*

Tributyltin oxide (TBTO) is a non-agricultural pesticide that is used in the shipping and boat industry in antifouling paints. These paints are used on marine boat hulls to minimise the growth of barnacles, algae and other invertebrates. This is a very real problem because on a long journey a large ship will accumulate many tons of plants and animals which introduce a significant drag factor and so increase the ship's fuel consumption. Bear this in mind when we discuss the environmental toxicity of TBTO in the next paragraph. It is very important to put this reduced fuel consumption on the benefit side of the risk:benefit equation.

It was noted recently that there was an increased proportion of males in dog whelk (*Nucella lapillus*) populations, but on closer examination it was demonstrated that some of the males were in fact females who had grown a penis; this phenomenon is termed imposex. After a very long search it was suggested that the reason for the dog whelks' imposex was their exposure to TBTO. This is a very serious ecological effect because it prevents the species from breeding and might result in its extinction.

Table 2.4 Toxicological effects of different concentrations of TBTO

TBTO concentration (μg dm^{-3})	Effect
0.001–0.01	Imposex in dog whelk
	Shell thickening in oysters
	Chronic toxicity in copepods
Concentration in harbour during TBTO paint removal = 0.06 μg dm^{-3}	
0.01–0.1	Growth inhibition in algae
	Growth effects in oysters
	Deformed limb regeneration in brittle stars
0.1–1.0	Acute toxicity in marine algae, oysters, copepods, mussels and mysids
	Reduced growth in fathead minnow
	Growth effect in trout
1.0–10.0	Acute toxicity in shrimp, *Daphnia*, sole larvae, trout and salmon
	Acute toxicity in ragworms

Data show that imposex in the dog whelk occurs at the ng dm^{-3} level and that environmental concentrations of TBTO in a harbour in the south of England were well within this range.

Some very elegant studies were carried out which demonstrated that the concentrations of TBTO in some harbours where ships' hulls were treated with antifouling paints were well within the range known to cause imposex (see Table 2.4).

Worries about the environmental impact of TBTO led to the Health and Safety Executive (UK Department of the Environment) to review the use of TBTO and in late 1995 it presented data to the Advisory Committee on Pesticides (a committee which advises ministers on pesticides) which led to significant restrictions in the use of TBTO by the maritime industry.

This example illustrates the speed at which the government can operate in response to potentially devastating environmental effects and supports the importance of the independent advisory committees in reviewing data. The problem, of course, is that the maritime industry will undoubtedly replace TBTO with something else; how long will it be used before we see its environmental impact?

It is important to assess the risk:benefit ratio of compounds such as TBTO. On the benefit side of the equation is the reduction in the drag of ships travelling long distances with a consequent reduction in fuel consumption. On the risk side is imposex in dog whelks and probably similar effects in other species. Clearly, here the equation fell firmly on the risk side and so TBTO was heavily restricted. The replacement for TBTO is likely to be copper which when incorporated into paints used on ships' hulls slowly dissolves in the sea so preventing the colonisation of invertebrates and algae. On the basis of our knowledge of copper toxicity in the marine environment it is a far safer alternative to TBTO, but with long-term use will the predictive power of our acute tests be proved right or wrong?

2.2.5 Organochlorine Pesticides and Bats

Bats are threatened by the effects of environmental pollutants and are now protected by law in many countries (e.g. in the UK this is under the Wildlife and Countryside Act (1981) and more recently under the Bonn Convention (1995)). In the 1980s the UK's Nature Conservancy Council became increasingly concerned about the effects of the use of remedial timber treatment products containing the organochlorine (OC) γ-hexachlorocyclohexane (γ-HCH, lindane) on the UK bat population. These products were used extensively to protect loft timbers against fungal rot and insect infestation (e.g. woodworm). Experiments confirmed that the use of γ-HCH at government-approved manufacturers' recommended application rates was capable of killing bats in domestic lofts. The UK government recognised this risk and placed restrictions on the use of γ-HCH-containing products. More recently (1996) they have introduced a universal classification and labelling scheme which denotes the effects of wood preservatives on bats in an attempt to minimise the impact of these chemicals upon this protected species.

2.2.6 Nitrates and Algal Blooms

Some deleterious effects of pollution have knock-on effects which have serious consequences. One such environmental impact relates to increased nitrate concentrations in surface water due to the indiscriminate use of nitrate-containing fertilisers. The nitrates are leached from farm land by rain, find their way into streams, then rivers, eventually entering lakes from where they have no route of escape. This has been happening for tens of years and plant growth has become a serious problem in some lakes. During the long and hot summer of 1986 in the UK this nitrification coincided with ideal light and temperature conditions for the growth of algae. Many lakes and ponds became green with algae, but some lakes (e.g. Rutland Water in the English Midlands) supported the growth of a specific genus of blue-green algae (*Mycocystis* spp.) which synthesise and secrete into their aquatic environment a potent toxin, mycocystin. Mycocystin has a complex toxicological profile: it is an acute hepatotoxin which at high doses results in death by liver failure, long-term exposure is neurotoxic.

Sheep and dogs died as a result of consuming water from *Mycocystis*-infested waters and the then Water Authorities became very worried about the presence of mycocystin in drinking water. The problem with drinking water was taken so seriously that some Authorities installed carbon filters to remove the toxin from water before it was piped to the consumer.

The seriousness of this example we hope will never be seen again, but each year algal blooms occur and toxins in water are monitored. This illustrates very well the problems associated with the coincidence of environmental conditions supporting the 'unnatural' growth of a member of a normally well-balanced ecosystem. There is a sinister postscript to this tale: during the hot summer of 1995, algal blooms occurred again and worries about mycocystin were resurrected, but this time we were ready.

2.2.7 Clenbuterol and Humans

It is important not to forget that we are members of the environment. Most environmental contaminants find their way into the human food chain and so in theory have some effect on us as the consumer. Very few food contaminants have any measurable effects; such effects are so rare that their occurrence usually is marked by a flurry of press activity, generally rather exaggerated (see Figure 2.8).

Clenbuterol, however, is an excellent example of a food contaminant which does have measurable effects in humans. Clenbuterol is a β-agonist (see Figure 2.9) and is used in veterinary medicine to treat respiratory disorders, particularly in horses. At very high doses it increases the proportion of lean meat to fat (this is called repartitioning) and therefore has an obvious use in the beef industry where there is a

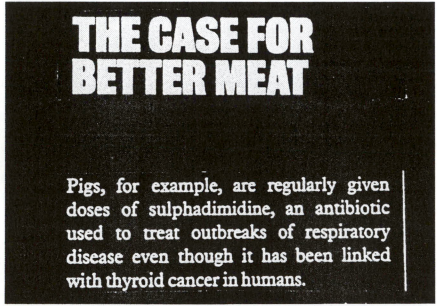

Figure 2.8 Newspaper headlines heralding food-related (or predicted!) effects of food chemical contaminants. In this case (as is often the case) there are errors which severely mislead the public. Sulphonamides are not *known* to be carcinogenic to humans, they are only carcinogenic in mice at very high doses. Newspaper articles should be read with a sceptical eye! Reproduced with permission of the *Sunday Times*, London.

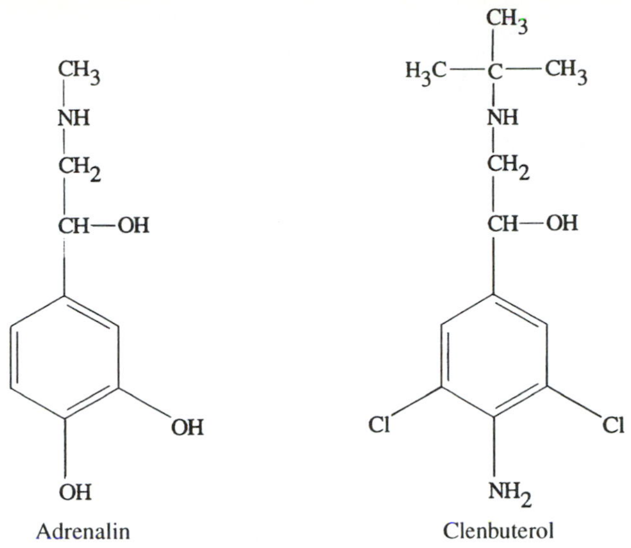

Figure 2.9 Molecular structure of the β-agonist clenbuterol (right) compared with adrenalin whose receptor it occupies in order to exert its pharmacological activity. The similarities are obvious.

price premium on lean carcasses. The problem is that clenbuterol is quite expensive and the doses necessary to result in economically significant repartitioning in beef cattle are so large that its use is economically non-viable. Besides economic reasons for its not being used in the beef industry, its use in meat-producing animals is illegal because hormones in meat production are banned in the EU (because clenbuterol occupies a hormone receptor it is classed as a hormone). Despite this, illicit cheap sources of clenbuterol became available in the mid-1980s and some farmers decided to risk its illegal use. Residues of clenbuterol in beef turned up in several countries in Europe (including the UK, Ireland and Spain). Residue levels in Spanish meat were associated with at least one human fatality (reported in the medical journal the *Lancet*). An enormous effort was undertaken to track down the source of the illicit clenbuterol in order to intercept it and so prevent its use because of the potentially deleterious effects that its residues in meat might have upon the consumer.

After a very great deal of detective work a most unusual association with illicit arms import was uncovered and clenbuterol's use declined and eventually disappeared. This is a good example of a food chain contaminant which would result in harm to the consumer.

2.2.8 Polychlorinated Biphenyls and Immunosuppression

Polychlorinated biphenyls (PCBs; see Figure 2.10) were commonly used insulating chemicals in the electricity industry. They are very toxic and were withdrawn from use several decades ago. Despite their withdrawal from commercial use their

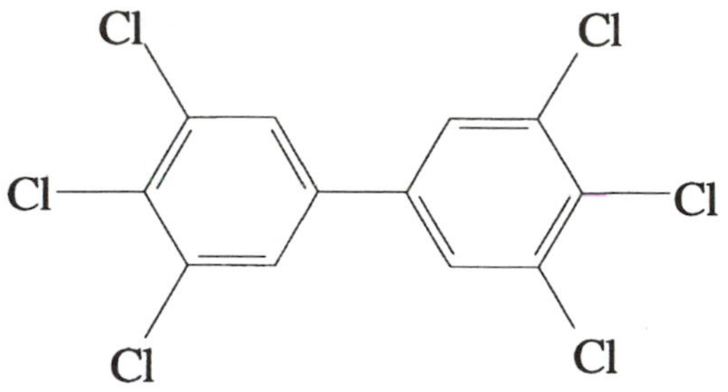

Figure 2.10 Molecular structure of a typical PCB (3,3',4,4',5,5'-hexachlorobiphenyl). Their environmental persistence is due to their stable hydrophobic structure.

environmental persistence means that they still remain in soil, water, plants and animals.

PCBs are blamed for many environmental ills, often without supporting scientific evidence. They are assumed to be the culprit. Despite our casting aspersions on their toxic effects they *are* immunosuppressive. This means that by mechanisms that we do not fully understand they inhibit an animals' immune responses and so reduce its ability to defend itself against bacterial and viral infections.

In the mid-1980s there was a sudden and unusual increase in the death rates of Arctic seals. It was unexplainable. Analysts found DDT, PCBs and many pesticide residues in the seals' tissues. This was hardly surprising because seals are very fatty creatures and so would be expected to accummulate these hydrophobic environmental pollutants. Some scientists concluded that the DDT and PCBs had caused the seals' deaths. Fortunately further studies demonstrated that the seals had died of canine distemper. This was a very strange finding indeed because it had not been seen in wild seals before. Seals are quite closely related to dogs and therefore it was not too surprising that if the conditions were right seals might contract the disease.

Further epidemiological studies showed that due to unusual circumstances the seals had come into contact with a feral colony of distemper-infected dogs. The high seal body burden of PCBs might well have immunocompromised the seals so preventing their confronting the virus and resulting in their deaths.

2.3   Risk versus Benefit

Life is a risky business! Crossing the road might result in death, flying in an aeroplane is to some a frightening experience because of the risk of a crash, having an operation in hospital, climbing mountains and so forth all have their attendant risks. All of these activities also have benefits and provided that the benefit outweighs the risk then we are prepared to accept the risk. For example, the risk of flying off to a holiday in a sunny resort in Spain is far outweighed by the benefit of a relaxing and happy holiday. The concept of risk, however, is more complex than this simple analogy. People are unable to assess risk accurately themselves and it is common for rarer activities to be associated with an apparently greater risk; this is termed perceived risk. Generally one is more worried about travelling by aeroplane than by car, this is because the perceived risk of air travel is greater. In fact the risk of being killed on the road is greater than the risk of being killed in an aeroplane accident. In assessing risk in scientific terms we must iron out these inconsistencies. We do this with the help of mathematicians who are able to define risk in strict mathematical terms.

But what is risk? Before we answer this question it is important to distinguish between hazard and risk. Hazard is an intrinsic property of a substance or an activity. For example, sodium cyanide is exceptionally hazardous because a small quantity ingested will cause death; however, the sodium cyanide must be ingested at a sufficiently high dose before death will occur, therefore the chance of ingesting (or coming into contact with) the cyanide must be part of the risk calculation. In fact:

RISK = HAZARD × CHANCE (OF EXPOSURE)

So, we have defined risk mathematically and removed the element of perception. To explore this further we can consider three hungry lions in three cages. The first is in a steel-barred cage with the door firmly bolted and locked. The second is in an identical cage with no lock or bolt, but the door is closed. The last is in a cage with the door open. The lions all have the same hazard (i.e. if they catch you they will kill you), however the chance of the first lion getting at you is very small and consequently the risk associated with walking in front of the lion's cage is correspondingly small. The risk associated with walking in front of the second lion's cage is greater because the chance of this lion escaping is greater because the door is not bolted and locked. The chance of the third lion escaping is very great indeed because its cage door is open – no sane person would walk in front of this lion because the risk of so doing is unacceptably high.

The concept of hazard and risk has been accepted for ever, but the idea of risk associated with exposure to chemicals is more recent. The idea that risk might be acceptable in the context of benefit is much more recent. Acceptance of risk formed the basis of the science of toxicology and occupies the attention of many regulatory committees (e.g. the Committee on Safety of Medicines (CSM)). You might, however, be surprised to learn that the concepts of hazard, risk and benefit were first aired by the German scientist Paracelsus in his Treatise published in the early sixteenth century.

Alle ding sind Gift	*All things are poisons*
und nichts ohn Gift	*There is nothing which is not a poison*
allien die Dosis macht	*It is the dose*
das ein Ding	*Which makes a thing safe*
kein Gift ist	

Paracelsus
1493–1541

Benefit is a rather more nebulus concept, but can perhaps be better appreciated by considering a medicinal example. If a new medicine were designed for the treatment of the common cold which had a 20 per cent mortality associated with its use, clearly the risk associated with its use is unacceptable in the context of a non-life-threatening disease. On the other hand, if the medicine were to be used for the treatment of cancer, even if the chance of a treatment success were only 30 per cent, it is likely that most people would accept the risk of medication because the cancer would kill if left unchecked. Here we have similar risks with very different benefits which change our acceptance of the risks.

In environmental terms hazard and risk are well understood and accepted, but benefit is far more difficult to understand and accept. One person's benefit is of no consequence to another. Environmental chemicals do not only pose personal risk and therefore no one person can unequivocally and uncontroversially determine the risk versus benefit. Again an example might help to clarify this point. Consider the use of nitrate as a fertiliser by farmers. The benefit to the farmer is great: his crops will grow better and he will make more money; from the point of view of a trout in a nearby brook there is no benefit, but an enormous risk of its habitat being destroyed by overgrowth of plants due to eutrophication. An honest broker has to be brought in to arbitrate and consider the global risk versus benefit – this person is the environmental toxicologist.

2.4 Development of Safer Chemicals

As we accrued experience of the adverse effects of pesticides upon the environment and became more concerned about the well-being of other animals and plants with which we share this finely balanced biosphere, the impetus to develop safer pesticides was born.

The OCs were well known to be dangerous to the environment because of their indiscriminate toxicity to (potentially) all animals. The swell of public opinion against the OCs, fired by Rachel Carson's *Silent Spring*, published in 1961 (see Chapter 9), resulted in several OCs being withdrawn; not even the companies making profit from the sale of these chemicals were able to resist. Clearly removal of OCs without their replacement with an effective, but environmentally more friendly, alternative would have been devastating to the agricultural industry. This provided the OPs with a very important role; they are less persistent (their $t_{1/2}$s in soil are measured in days rather than years) and so were ecologically more acceptable insecticides than the OCs. They soon replaced the OCs for most insecticidal applications in many parts of the world.

People were not satisfied that OPs were safe enough. As discussed above there is a great deal of evidence that even the OPs are environmentally unacceptable in the long term. The search was therefore on for even safer pesticides. There was a financial advantage for companies producing 'environmentally friendly' pesticides because the swell of public opinion in favour of environmental protection meant that if such pesticides could be developed they would be quickly accepted even if their price were greater.

2.4.1 *Pyrethroid Insecticides*

The pyrethroids are often regarded as a modern generation of 'safe' insecticides: far from it! They were used by the Chinese about 1000 BC in the form of ground or powdered chrysanthemum leaves to deter insect infestation of ornamental plants. Pyrethrum is a mixture of several pyrethroids present in powdered *Chrysanthemum* (*Pyrethrum*) *cinerariaefolium* or *C. coccineum*, including pyrethrin, pyretol, pyrethrotoxic acid, pyrethrosin and chrysanthemine. The remarkable insecticidal properties of these pyrethroids possibly explains why chrysanthemums are relatively resistant to insect attack. The most common insect pest to bother chrysanthemum growers is the leaf miner, a tiny insect larva which mines just below the surface of the leaf; presumably this creature has evolved a resistance to the pyrethroids which allows it to occupy this unique ecological niche.

The use of pyrethroid mixtures as insecticides began in 1850. In 1965 the total world production of pyrethroids from natural sources was 20 000 tons; some 50 per cent of this was produced in Kenya where *C. cinerariaefolium* grows well. Until quite recently the pyrethroid insecticides were only isolated from plants because their molecular structure was very complex (see Figure 2.11) and made their chemical synthesis very difficult. They are now routinely synthesised by the agrochemicals industry. An important fairly recently introduced pyrethroid is cypermethrin (see Figure 2.11), once its price comes down it is likely to be a relatively environmentally friendly alternative to the OPs.

Cypermethrin is a synthetic *cis/trans* pyrethroid insecticide commonly used as an ectoparasiticide in veterinary medicine. It occurs as *cis* and *trans* isomers, the former being a more potent insecticide.

The pyrethroids (including cypermethrin) are thought to act by modulating the gating characteristics of the sodium channel on the neuronal membrane although the exact mechanism of the interaction between the pyrethroid molecule and the membrane sodium channel is not fully understood.

The pyrethroids in general are very rapidly metabolised (see Figure 2.12) in mammals and ecosystems which possibly explains their low acute and chronic toxicity; they are, however, selectively more toxic to fish which may reflect species differences in metabolism. It is well known that the rate and extent of metabolism has a significant effect upon their toxicity, presumably because only the parent pyrethroid is active in blocking the membrane sodium channel.

The metabolism of the pyrethroids (see Figure 2.12) involves esterase cleavage of the central ester bond, cytochrome P_{450}-catalysed oxidation at several positions on the molecule (e.g. the methyl group attached to the cyclopropane moiety) and phase II conjugation of the products of these phase I metabolic pathways with glycine, glucuronic and sulphuric acids. It is an extensive metabolic pathway with many facets which are prone to species variations and for this reason one might expect significant species differences in the metabolism of cypermethrin. Because of the rapid loss of activity of cypermethrin in ecosystems (i.e. when used in the field) it is very likely that only the parent compound has insecticidal properties and therefore in the context of residue studies it is likely that only the parent compound

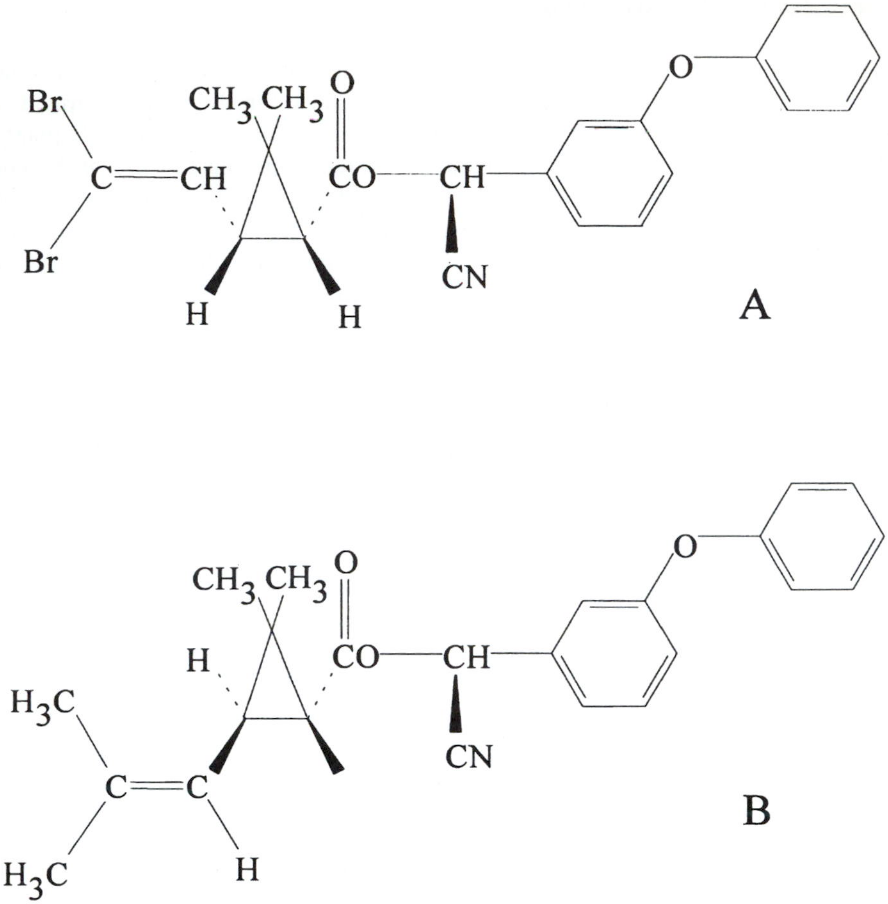

Figure 2.11 Molecular structures of Decamethrin (A), one of the first pyrethroid insecticides to be synthesised, and Cypermethrin (B), a modern insecticide which can be used to replace OPs in sheep dips and as a crop insecticide.

is of toxicological significance. Interestingly, the pyrethroids are profoundly selective for insects, having very low mammalian toxicity; this low toxicity in mammals suggests that as there is extensive metabolism in mammals the metabolites are not significantly toxic. Indeed the metabolic pathway results in a significant increase in the water solubility of the pyrethroids so facilitating their rapid elimination from the body in the water-based excretory fluids (e.g. urine).

The pyrethroids are therefore, despite their extremely potent insecticidal activity, relatively environmentally safe because they are very rapidly degraded by UV-light-catalysed photolysis and microorganism (and most other species) metabolism. It is interesting to reflect that nature 'designed' these insecticides and clearly it was not in her own interest to design a chemical which would be

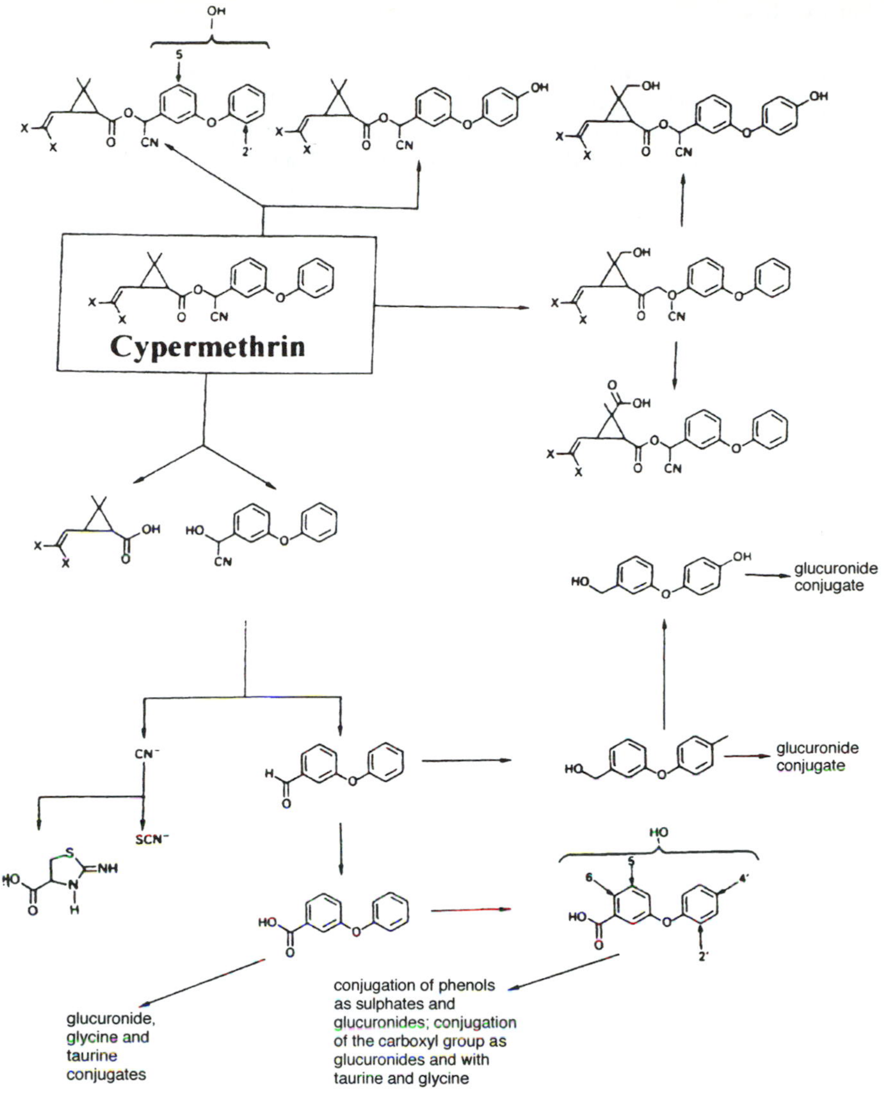

Figure 2.12 Generalised metabolic pathway for Cypermethrin. Elements of the pathway occur in mammals, insects and plants; the pathway is, however, based on the metabolism in mammals. Adapted from Cremlyn (1990), *Agrochemicals*, Wiley.

persistent and so put her own future in jeopardy. In commercial terms the rapid degradation of pyrethroids (and therefore rapid loss of activity) is to some extent undesirable and therefore it is likely that we will alter their molecular structures to optimise their activity/degradation profile. We must be careful though, otherwise we will reduce their environmentally friendly attributes.

Further Reading

Calow, P. (Ed.), 1993, *Handbook of Ecotoxicology*, Vol. 1, Oxford: Blackwell.
Environmental Health Criteria 63, 1986, *Organophosphorus Insecticides*, Geneva: WHO.
Leahey, J.P. (Ed.), 1985, *The Pyrethroid Insecticides*, London: Taylor & Francis.
Turnbull, A., 1997, Chlorinated pesticides, in Hester, R.E. and Harrison, R.M. *Chlorinated Organic Micropollutants*, London: Royal Society of Chemistry.

3

Environmental Toxicity Testing

The philosophy and practicalities of environmental toxicity testing are discussed. The trophic level approach to testing is focused on and critically evaluated. The process of testing and the test species used are outlined.

Quick toxicity tests, oestrogenicity, physicochemical properties as a means of predicting environmental behaviour and structure-activity relationships are discussed.

3.1 Environmental Toxicity Testing in Perspective

Toxicologists have debated for many years the wisdom of extrapolating the results of toxicity tests in rats (and other species) to humans. The debate usually centres upon whether a particular species is a good model for people. The conclusion of such debates is generally that the only real test for toxicity in humans is to use ourselves in the toxicity test. This, of course, raises significant ethical problems and therefore a compromise has been struck whereby a battery of toxicity tests using different animal species is used for assessing the safety of a new drug and the tests are crowned with carefully controlled clinical trials when the toxicologists have no reason to believe that the drug will cause the volunteer or volunteer patient any ill effects. It is therefore necessary to finalise the toxicity testing of a new drug with a 'test' in the species for which the drug is intended. This rule is also true for veterinary medicines, although in this case the ethical situation is different and it is considered appropriate to test the drug in the target species very much earlier in its development programme than is the case for human medicines.

At the end of the testing period the company developing the drug present an Approval for Marketing Application to the Department of Health (DoH) via the Medicines Licensing Agency (MLA) for human medicines or the Ministry of Agriculture, Fisheries and Food (MAFF) via the Veterinary Medicines Directorate (VMD) for animal medicines. The package submitted includes an extensive assessment of the safety of the drug. If the dossier is approved and the drug is licensed for marketing it is at this point that it becomes a medicine. (A drug is a pharmacologically active substance, a medicine is used to treat disease; see Chapter 5 for a detailed discussion of toxicity tests and Chapter 8 for details of the legislation involved.)

Pesticides approval is similar. It too involves a committee (the Advisory Committee on Pesticides; see Chapter 8) considering a toxicology package before approval for marketing can be granted.

The approval systems outlined above are specific to the UK; however, very similar approaches are used by most developed countries. Indeed they are often based upon the British system which was one of the first to be introduced (see Chapter 8).

The need to assess the risk of medicines and pesticides to the environment is assuming very great importance. The need is greatest for pesticides because they are often applied directly to the environment (e.g. insecticides sprayed onto crops) and undoubtedly have an effect upon the ecosystems which they contaminate. Veterinary medicines are next in order of environmental impact importance because they are applied to (e.g. ectoparasiticides such as sheep dips) or given to (e.g. anthelmintics) animals which then drip the active ingredients of the medicine (e.g. OPs in sheep dips) or excrete the medicine and its metabolites (e.g. ivermectin in some anthelmintics) into the environment. The amounts here are very much smaller than for pesticides which are applied directly to fields and are actually designed to kill the fields' natural inhabitants.

At the bottom of the priority list, with respect to risk to the environment, are human medicines. When we take a medicine we excrete it and its metabolites in our urine and faeces and they might find their way into the environment via the sewage treatment works. A great deal of work has been carried out, particularly in the anticancer drug field, to investigate the potential for environmental contamination by human medicines. Generally, the microorganisms in the sewage treatment works solve the problem by degrading the drugs and their metabolites. The risk to the environment is therefore small.

As we become more concerned about the effects of human activities upon the environment we have appreciated the need to predict the potential effects of pesticides and medicines upon ecosystems. It is now a requirement of the licensing authorities that the safety data package includes environmental toxicity testing and that an environmental impact assessment is made.

3.2   The Scope and Limitations of Environmental Toxicity Testing

We discussed above (see Section 3.1) the heated arguments which surround extrapolation of animal toxicity data to humans in assessing the risk of medicines. This pales into insignificance when environmental toxicity is considered. When we consider human medicines we are concerned about the well-being of one species, namely humans. When we consider environmental toxicity we are concerned about the well-being of 10^7 species ranging from single celled organisms to massive complex animals such as the blue whale (*Balaenoptera musculus*); of course, we must not forget humans here, because they too are part of the environment. This is rather ironic; humans are producing chemicals which might have deleterious effects upon their own ecosystem and so adversely effect them.

The need to assess the effects of pesticides and medicines (and industrial chemicals, although these do not fall within a single legislative structure) upon a vast array of different animals and plants is an almost insurmountable problem. It is obviously not possible to carry out target species testing. Indeed, this, in most cases, would be undesirable because the target species are often endangered (e.g. eagles) and we would use more individuals in our toxicity testing regimens than might be affected by the chemical when it is used. It is for these reasons that the philosophy of environmental toxicity testing is different to that for human risk assessment.

3.3   The Philosophy of Environmental Toxicity Testing

Instead of target species toxicity testing environmental risk assessment relies upon target trophic level (see Chapter 1) toxicity testing. Representative species of the trophic levels are subjected to toxicity testing and the results extrapolated to other members of the same trophic level. There is complete agreement among toxicologists about the validity of this approach; it is not valid, but what else can we do? It is possible to illustrate this lack of validity by one extreme example. Trophic level 2 (primary consumers) is represented by *Daphnia magna* in environmental toxicity testing. *Daphnia* (see Figure 3.1) are small aquatic animals which are easy to breed in the laboratory and therefore lend themselves very well to toxicity testing. At the other end of the spectrum, the hippopotamus (*Hippopotamus amphibius*) is also in trophic level 2. We are therefore using *Daphnia* to represent the hippopotamus. Clearly this is ridiculous. It is important that the risk assessor keeps such extreme examples in mind when deliberating upon environmental toxicity testing data.

We will return to the trophic level approach to toxicity testing later (see Chapter 5). From the discussion so far it is clear that toxicity testing alone is fraught with problems; for this reason other properties of the molecule being studied are used in the final assessment. If all of the properties and the toxicity testing point to the same conclusion the environmental toxicologist can be more confident in drawing a particular conclusion; if not he might have to re-think.

The physicochemical properties (see Chapter 6 for a discussion of the importance of physicochemical properties) of the molecule and its structural

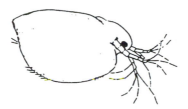

Figure 3.1   Water flea (*Daphnia magna*) an important test organism in ecotoxicity testing. It is about 2 mm long.

analogy to chemicals with known environmental effects (structure-activity relationships, SARs) are powerful predictive tools and form an important facet of the environmental impact or risk assessment.

3.4 The Process of Environmental Risk Assessment

3.4.1 *Physicochemical Properties*

One of the most important properties of a molecule in conferring environmental toxicity is its propensity to concentrate up the food chain. If a molecule is soluble in fat it will reside in the biological membrane which surrounds all cells (see Chapter 1) and when its 'host' is eaten the molecule will seek out the consumer's biological membranes and once again form long-term residues there. This process continues up the food chain resulting in concentration and increasing the potential for animals at the top of the food chain being adversely affected by the chemical concerned (e.g. raptors and DDT, see Section 2.2.1). Predicting the potential of a new medicine or pesticide to seek out cell lipids is a very important part of the environmental risk assessment strategy. Physicochemical properties are discussed fully in Chapter 6.

3.4.2 *Structure-Activity Relationships*

Particular chemical groups confer specific properties to molecules. For example, if a molecule has a carboxyl group (-COOH) it will have acidic properties. At the toxicological level it is well known that some groups (or moieties as they are often called) are associated with particular toxicological traits. For example, aldehyde moieties (-CHO) are associated with immunological sensitivity in mammals. The mechanism of this example is that a small aldehyde-containing molecule (i.e. too small to illicit an immunological response on its own) can react with the terminal amino group of a lysine residue in a protein to form Schiff's base ($-C(O)-NH-CH_2-$). This means that the protein has been modified and appears foreign to the host mammal (i.e. it is a hapten–carrier complex). The animal's immune system generates antibodies directed against the foreign chemical. If the mammal is re-exposed to the chemical it will immediately be sensitive and might initiate a severe immunological response (i.e. the process of sensitisation).

There are many more examples of SARs, particularly in the field of carcinogenesis. SARs can be very specific, not only relying upon a particular moiety being part of the molecule's structure, but requiring it to be in a specific position or conformation on the molecule. This is illustrated well by considering the aromatic amino group and carcinogenesis (see Figure 3.2). Here the amino group must be at the end of a long planar (i.e. flat) aromatic structure (e.g. benzidine, see Figure 3.2) in order to direct it into the correct position of the DNA molecule where it can initiate damage and result in carcinogenesis.

Benzidine is one of the most carcinogenic substances known, indeed its use in the UK is banned. It is a planar molecule with two aromatic amino groups.

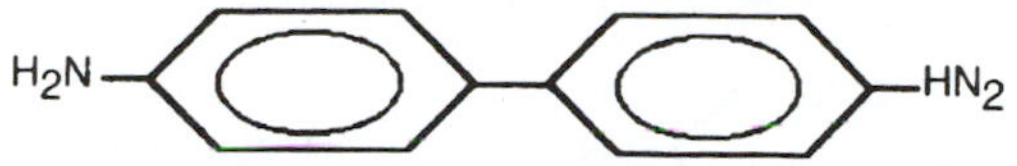

Benzidine

This molecular planarity facilitates intercollation into DNA and the positions of the amino groups possibly allow a specific interaction with a DNA base (probably guanine) which results in irreversible DNA damage and eventually carcinogenesis. Benzidine is a genotoxic carcinogen.

Based on this philosophy one might predict aniline to be a potent carcinogen too.

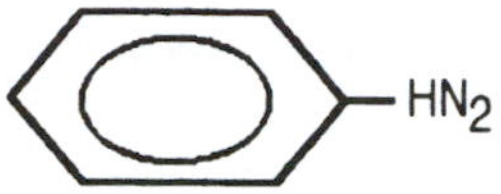

Aniline

However, it is not. The SAR for aromatic amino compound carcinogenesis therefore requires a planar hydrophobic aromatic region of significant size (e.g., two benzene rings) plus one, or perhaps two, amino groups.

We can test this predictive hypothesis using 2,4' biphenyldiamine (2,4-BD).

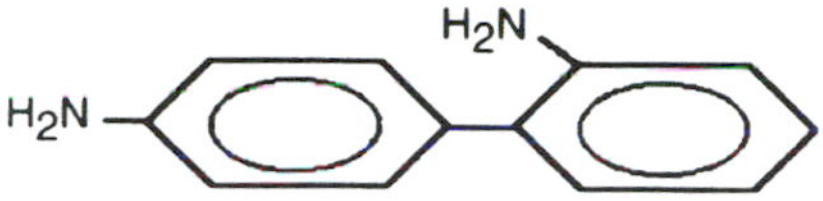

2,4'-Biphenyldiamine

Wrong! 2,4'-BD is not a carcinogen. Our hypothesis must be modified: perhaps the amino groups have to be in the 4-position. This can be tested using biphenylamine (BPA).

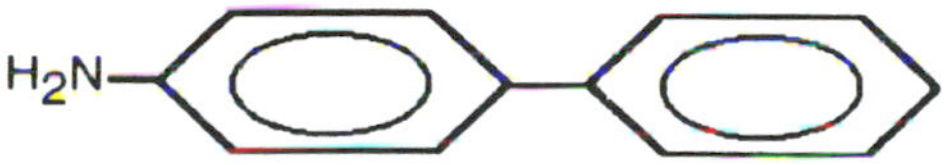

Biphenylamine

Correct! BPN is carcinogenic, but not as potent a carcinogen as benzidine.

It appears that if a non-4-position amino group is present that this prevents a planar aromatic amino group from being carcinogenic. This is an important predictive property and therefore the hypothesis must be tested further. The planar aromatic amino dye, chrysodine, proves the point nicely because it is not carcinogenic.

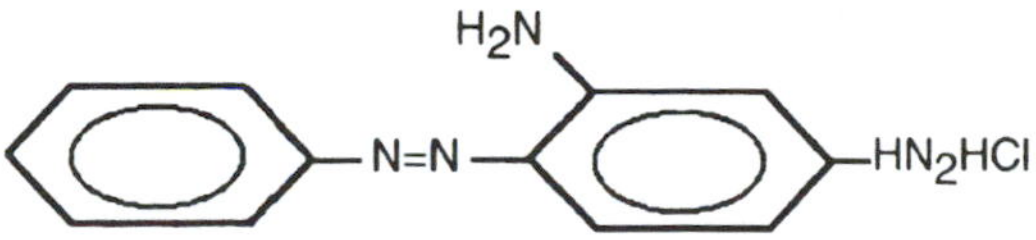

Chrisoidine

Figure 3.2 Structure-activity relationships for aromatic amino and nitro compounds and carcinogenesis showing the importance of the position of the groups in conferring carcinogenicity. From I.C. Shaw and J. Chadwick, 1995, *TEN*, **2**, 84.

We are now beginning to build a good predictive picture in relation to carcino-genicity and planar aromatic amines. This line of reasoning can be continued to include other chemical moieties which have 'similar' properties to the amino group (e.g., the nitro group). Indeed 4-nitrobiphenyl is carcinogenic,

4-Nitrobiphenyl

whereas 2-nitrobiphenyl, i.e.

2-Nitrobiphenyl

is not. So the same rules seem to apply. If the rules are applied to the herbicide, nitrofen,

Nitrofen

the predictive system developed thus far would alert one to nitrofen being a potential carcinogen. Indeed the IACR categorise nitrofen as 'reasonably anticipated to be a carcinogenic'. The system therefore works!

Figure 3.2 *cont'd.*

The above examples are not specific environmental effects, but of course they could both occur in many species and so have environmental significance. Environmental toxicologists are only now beginning to develop SARs for environmental effects. Perhaps the best example is for the environmental xeno-oestrogens.

The oestrogenic effects of chemicals in the environment are potentially very serious indeed. We have witnessed alligators (*Alligator mississippiensis*) in the Everglades in Florida in the USA losing their penises, female dog whelks (*Nucella lapillus*) growing penises and men experiencing a significant reduction in sperm count. The effects of these phenomena upon reproductive capacity and hence survival of the species could be very severe indeed. It is important that we are able to predict if a new molecule (e.g. a pesticide) might have oestrogenic activity.

The mechanism of environmental oestrogenicity involves the putative oestrogen fitting the oestrogen receptor (ER) in the host animal. If the molecule fits and has the appropriate molecular attributes to activate the receptor it will 'fool' the receptor into seeing it as an oestrogen (e.g. 17β-oestradiol) and therefore ellicit the oestrogen response (i.e. feminisation). A detailed knowledge of the ER will allow the toxicologist to determine which facets of a 'real' oestrogen are necessary to 'switch on' the receptor. This work is in its infancy, but it is clear that regions of electronegativity (e.g. hydroxyl groups) separated by a long hydrophobic region are necessary (see Figure 3.3). Molecules with molecular configurations similar to

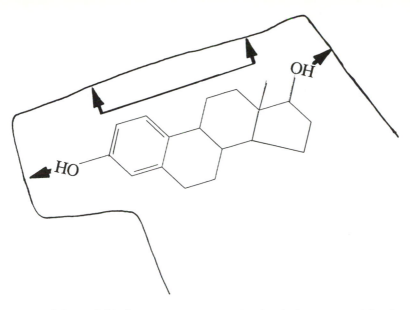

Figure 3.3 17β-Oestradiol in the oestrogen receptor showing the importance of the phenolic hydroxyl groups and the long hydrophobic region. Adapted from Müller *et al.*, 1995, *TEN*, **3**, 70.

17β-oestradiol are likely to fit and activate the ER. There are many examples of known xeno-oestrogens (unnatural or foreign oestrogens — xeno is from the Greek for foreign) which can be used to test this hypothesis (see Figure 3.4).

Computer programmes are being developed currently which will allow a new molecule to be scrutinised to determine whether its molecular structure and configuration might facilitate its activating the ER. These will be a significant aid to the environmental toxicologist, but it is important that they are not used blindly because idiosynchratic xeno-oestrogens might be missed.

3.4.3 *Oestrogenicity Assays*

3.4.3.1 In Vitro *Assays*

As discussed above oestrogenicity depends upon a molecule fitting and activating the oestrogen receptor. There are now oestrogenicity assays that utilise yeast cells which express the human oestrogen receptor (hER). The receptor is coupled to the enzyme galactosidase so that when the hER is occupied by an oestrogenic compound galactosidase is produced. A yellow galactoside dye is included in the yeast culture medium and when an oestrogen is present the galactosidase produced cleaves the galactoside unit releasing the chromophore which becomes red in the free state (see Figure 3.5). The yeast therefore become red in the presence of an oestrogen (see Figure 3.6).

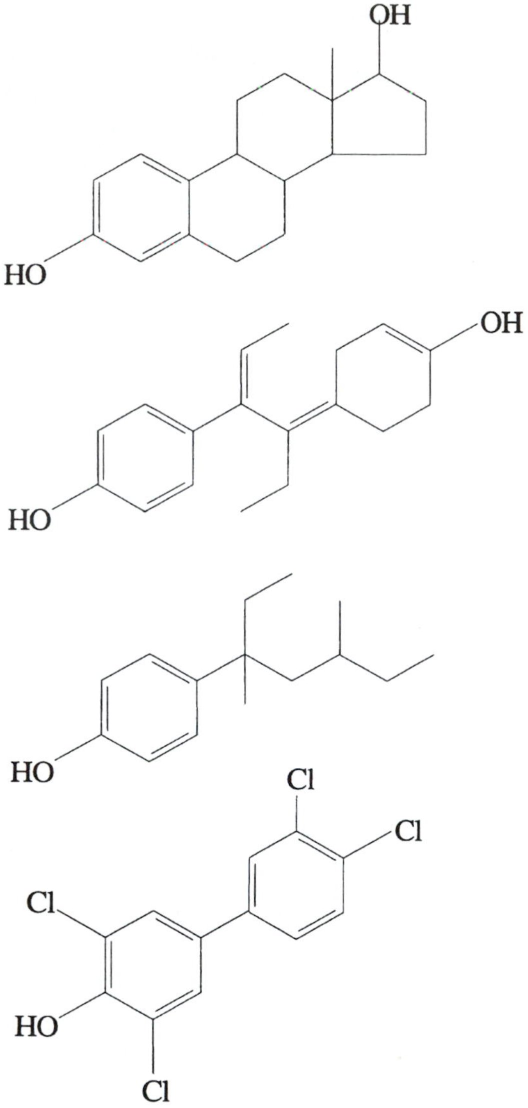

17β-oestradiol

Diethylstilboestrol

An isomer of nonylphenol

A polychlorinated biphenyl
(PCB) congener

Figure 3.4 Examples of xeno-oestrogens; their structural analogy to oestradiol can be seen
clearly. Adapted from Müller *et al.*, 1995, *TEN,* **3**, 69

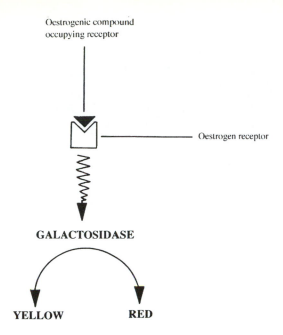

Oestrogenic compound occupying receptor

Oestrogen receptor

GALACTOSIDASE

YELLOW   **RED**

Figure 3.5 Schematic representation of the yeast oestrogenicity assay showing the oestrogen receptor occupied by an oestrogenic compound with consequent activation of galactosidase. The result is yellow galactoside dye being cleaved to release the red chromophore.

3.4.3.2 In Vivo *Tests*

Male fish produce the egg protein vitellogenin (an egg protein normally produced only in the female) when exposed to oestrogens. The protein is reasonably easy to measure and forms the basis of a bioassay for oestrogenicity. Male fish (rainbow trout (*Salmo gairdneri*) or zebra danio (*Brachidanio rerio*)) are exposed to suspected oestrogens in their water and after a considerable period of exposure (weeks or months) vitellogenin is measured in their tissues. A positive analysis for vitellogenin signifies the presence of an oestrogen.

3.4.4 *Toxicity Tests on Animals and Plants*

As discussed previously environmental toxicity testing relies upon a trophic level approach using representatives of each trophic level in specific toxicity tests (see Table 3.1). The tests concentrate on a freshwater aquatic ecosystem; however, individual tests on terrestrial and marine organisms are often carried out to give a better indication of the broader environmental impact of the test chemical.

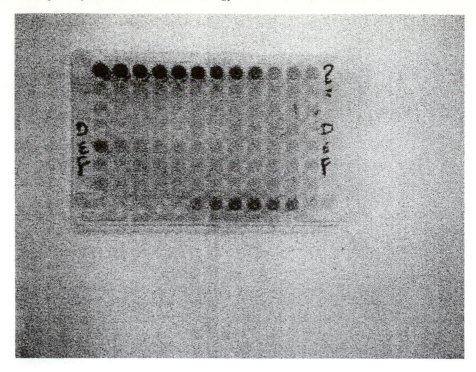

Figure 3.6 Typical oestrogenicity assay using a yeast expressing the human oestrogen receptor. The dark wells (actually red) signify oestrogenicity. *Top row:* A series of 17β-oestradiol standards, highest concentration on left. *5th row from top* (labelled E): An experiment to assess the oestrogenic potential of the pesticide Permethrin showing very weak activity at very high concentrations (highest concentration on left) of the pesticide. *Bottom row:* Tamoxifen (an oestrogen analogue used in the treatment of breast cancer), highest concentration on the right. This is used as a positive control to test whether the system is working. From experiments carried out by Dr Colin Davy then in the Centre for Toxicology, University of Central Lancashire, UK.

3.4.4.1 *Trophic Level 1*

The common freshwater algae *Selenastrium capricornatum*, *Scendesmus subspicatus* and *Chlorella vulgaris* are often used to represent trophic level 1 because they grow rapidly and are easy to culture. The test relies upon biomass production over time (usually 2 days) in the presence of increasing concentrations of the test chemical. The concentration of test chemical necessary to inhibit the growth of the algae by 50 per cent is determined (i.e. the EC_{50}: effective concentration for 50% reduction in growth).

The test is simple. Algae are inoculated into culture vessels containing increasing concentrations of the test chemical (plus a control with no added test chemical). The cultures are incubated in an illuminated incubator for 2 days. At the

Table 3.1 Trophic levels showing the commonly used representatives in ecotoxicity testing

Trophic level	Description	Example	Species generally used in testing
1	Primary producer	Algae	Blue green algae
2	Primary consumers	*Daphnia*	*Daphnia magna*
3	Secondary consumers (carnivores)	Water spider	No defined species
4	Tertiary consumers (carnivores)	Fish	Trout/blue gilled sun fish
5	Quarternary consumers (carnivores)	Birds of prey	Quail[a]

[a] Not a typical level 5 consumer but used to represent the type because of ease of keeping; however, this animal does represent a human food species which is another important consideration.
From Shaw and Chadwick, 1995, *TEN*, **2**, 81.

end of the incubation period the growth of the algae is determined by filtering out the algae using a pre-weighed filter, re-weighing the filter plus algae and so determining the biomass of the algae produced during the culture period. The algal biomass is plotted against the test chemical concentration and the concentration at which the biomass is 50 per cent of the control value is determined; this is the EC_{50} (see Figure 3.7).

The EC_{50} value is very useful because it enables the environmental toxicologist to assess the concentration of the test chemical which will have a deleterious effect upon the environment. For example, if the $EC_{50} = 1.2 \, \text{mg dm}^{-3}$ and the predicted

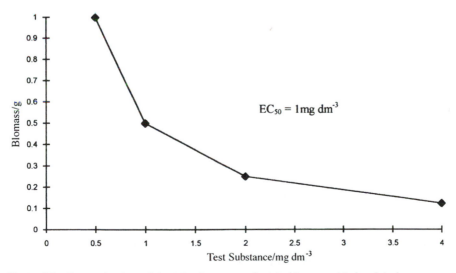

Figure 3.7 Determination of the EC_{50} for a test chemical in a trophic level 1 algae test.

concentration of a particular chemical in a river (e.g. this might be a pesticide used on a field next to a river), is $0.001\,mg\,dm^{-3}$, the impact of the test chemical upon the river is likely to be negligible because the worst case predicted concentration in the river is very much lower than the EC_{50}. The worst case predicted concentration in the river would be the total amount of pesticide applied to the field divided by the approximate volume of the river adjacent to the field. This is the worst case because it does not take account of the river's flow and assumes that all of the pesticide will enter the river. Environmental toxicologists often use worst case scenarios because the 'real' situation can only be better.

3.4.4.2 Higher Plant Tests

Higher terrestrial plant tests are often carried out. Food plants (see Table 3.2) are generally used to assess the effects of the test chemical upon the growth of crops. These tests have two functions: primarily to assess the effect of the test chemical upon the growth of the test plant, but also to allow residues of the test chemical in the plant to be determined. The latter data are used to assess the risk of the test chemical (e.g. a pesticide) to the consumer of the crop.

Three parameters are usually measured: namely germination frequency, root elongation and shoot elongation. The results are usually expressed as the concentration of the test chemical at which growth or germination retardation is evident.

Table 3.2 Three categories of plants used in the terrestrial plant growth test based on the major human food plant groups

Category 1	Monocotyledons e.g. rye grass, rice, oats, wheat, sorghum
Category 2	Brassicas e.g. mustard, rape, radish, turnip, Chinese cabbage
Category 3	Legumes, etc. e.g. vetch, mung beans, red clover, fenugreek, lettuce, cress

From Shaw and Chadwick, 1995, *TEN*, **2**, 82.

3.4.4.3 Trophic Level 2

For freshwater aquatic ecosystems the water flea (*Daphnia magna*; see Figure 3.1) is almost always used and for marine systems the brine shrimp (*Artemia salina*) is often used. Both of these species are easily cultured in the laboratory. *Artemia* is particularly interesting and well suited to laboratory culture because its eggs will withstand drying and storage for long periods of time. When the dried eggs are sprinkled onto sea water and incubated at about 23°C they hatch within a day or two and are ready for use in ecotoxicity testing.

The test is carried out in much the same way as the trophic level 1 test; an EC_{50} is the outcome. In the trophic level 2 test the parameter measured is not growth (i.e. biomass), but rather the effect of the test chemical upon the individual test animal.

Culture vessels (beakers containing aerated fresh water for *Daphnia* or sea water for *Artemia*) are prepared with culture medium containing increasing concentrations of the test chemical (plus a control with no test chemical). To the vessels are added 10 test organisms. They are incubated in the light for a specific time. At the end of the incubation period the number of immobilised organisms is determined. It is difficult to decide whether *Daphnia* or *Artemia* are dead which is why immobility is used as the endpoint. The test is usually carried out in triplicate because a small number of test organisms are used per test chemical concentration, which allows for statistical variation in immobilisation to be accounted for.

The EC_{50} is calculated by plotting the proportion (usually as percentage of total) of test animals immobilised against the concentration of the test chemical. The calculation is the same as that for the algae EC_{50} (see Figure 3.7).

The *Daphnia* test is divided into two parts:

1. *Acute EC_{50} test*: *Daphnia* are exposed to the test chemical for 24 h and immobilisation determined.

2. *Reproduction test*: *Daphnia* are exposed to the test chemical for 14 days. The increase in number of *Daphnia* is counted. The effect of the test chemical upon reproduction is thus determined.

3.4.4.4   *Trophic Level 3*

Representatives of trophic level 3 are not included in the environmental toxicity testing regimen. This is because their biochemistry and physiology and therefore presumably their response to toxic chemicals is similar to animals from trophic level 4. This is a sweeping statement because the diversity of species in trophic level 3 (as in all of the trophic levels) is enormous, but if specific animals from trophic level 3 are compared with similar species from trophic level 4 the point is made. For example, the *Daphnia*-eating water spiders (*Araneida* spp.) are in trophic level 3, whereas *Portia* spp., the jumping spiders from southeast Asia which eat other spiders, are in trophic level 4. As they are both spiders it is likely that their response to toxic chemicals will be similar.

3.4.4.5   *Trophic Level 4*

Trophic level 4 is represented by fish. The fish used most commonly by the environmental toxicologist are the rainbow trout (*Salmo gairdneri*, a very sensitive species which has exacting requirements of water purity and high oxygenation) or the blue gilled sunfish (*Lepomis macrochirus*, a North American fish of lakes and slow flowing rivers); however, many other fish species are also used (see Table 3.3). The fish are either exposed to the test chemical in their water at varying concentrations and mortality is monitored and an LC_{50} (the concentration that kills 50 per cent of the population) determined, or the fish are dosed individually with the test chemical, either by injection or by incorporation into pelleted feed. Mortality is again monitored, but this time an LD_{50} (the dose necessary to kill 50 per cent of a population) is determined. By far the commonest procedure used is the LC_{50} test.

Table 3.3 Fish species used in ecotoxicity testing

Species	Facet of the environment[a]
Fathead minnow (*Pimephales promelas*)	Estuarine
Zebra fish (*Brachidanio rerio*)	Lotic: warm/fresh water
Common carp (*Cyprinus carpio*)	Lentic: cold/fresh water
Guppy (*Lebistes reticulatus*)	Lentic: warm/fresh water
Rainbow trout (*Salmo gairdneri*)	Lotic: cold/fresh water
Blue gill sunfish (*Lepomis macrochirus*)	Lentic/lotic: cold/fresh water

[a] Lotic: flowing water; lentic: still water.
From Shaw and Chadwick, 1995, *TEN*, **2**, 83.

The acute test involves exposure to the test chemical for a total of 96 h. Fish mobility is recorded at 24 h intervals and a cumulative LC_{50} is calculated. The chronic toxicity test covers a 14-day period with continual exposure to the test chemical. Unfortunately these tests do not probe the aquatic secondary consumers fully because they do not take account of the dietary intake of the test chemical in the form of food chain-concentrated residues.

3.4.4.6 Trophic Level 5

The quarternary consumers are at the top of the food chain. In ecological terms they are the species which are most at risk from the deleterious effects of environmental pollutants. The problem is that they are often rare (perhaps due to previous human activity) and so it is unacceptable to use these species *per se* in ecotoxicity testing. For example the peregrine falcon (*Falco peregrinus*) is in trophic level 5, it is endangered and cannot be used in toxicity testing. This would be the case for most representatives of trophic level 5. For this reason a species from the same family which is not in trophic level 5 is used to represent this trophic level. Perhaps the most at risk group in trophic level 5 is the raptors (birds of prey). They are members of the family Aves (birds) and it is for this reason that the environmental toxicologist chooses a bird, often the bobwhite quail (*Colinus virginianus*) (see Table 3.4), to represent the trophic level. The quail is far from being a raptor (it is a timid bird), but its metabolism and therefore susceptibility to toxicity are possibly akin to that of its relatives, the raptors.

The test is carried out either by administering the test chemical to the birds in their diet or by injecting it into them (usually intraperitoneally, i.e. into the abdominal cavity). There are two ways of proceeding. The first is to use death as an endpoint to determine the LD_{50}. The second (and more humane) is to use an endpoint which points to the test chemical having had an effect upon the bird, either physiological (e.g. effect upon growth rate) or biochemical (e.g. effect upon a blood enzyme). In this case the NEL or NOEL (see Chapter 5) is determined. Both the NEL (or NOEL) and LD_{50} signify the dose at which the test chemical has deleterious effects on the test bird and are important in environmental impact assessment.

Table 3.4 Bird species used in the avian test to represent trophic level 5

Bobwhite quail	*Colinus virginianus*
Pigeon	*Columba* spp.
Ring-necked pheasant	*Phasianus colchicus*
Reg-legged partridge	*Alectoris ruffa*
Mallard duck	*Anas platyrhynchos*

In addition to the above acute toxicity tests it is also necessary to carry out multiple generation toxicty tests which are akin to reproductive toxicity tests in mammals (see Chapter 5). The mallard duck (*Anas platyrhynchos*) or quail are generally used for these studies. The birds are administered with the test chemical (usually in their feed) for at least 20 weeks during the egg laying season (this can be artificially created by controlling the day length with artificial light). The eggs are collected and incubated. The hatching 'efficiency' and viability of the chicks or ducklings are determined to ascertain whether the test chemical might affect the reproductive potential of birds if it pollutes the environment. In addition, the chicks are observed for a 14-day period. A NOEL based on such parameters as number of eggs laid, eggshell thickness and hatching mortality is determined. This is an important test because it ensures that we do not face another DDT/eggshell thinning problem and it brings us full circle, because it was the DDT problem which made us realise the importance of environmental toxicity testing.

3.4.4.7 *Terrestrial Toxicity Tests*

Most of the toxicity tests study aquatic ecosystems (only the avian test falls outside this generalisation); however, terrestrial ecology is considered both by extrapolating results from aquatic systems (assuming similar toxic susceptibilities between aquatic and terrestrial members of particular trophic levels) and by conducting experiments on terrestrial organisms. The most common of these tests is carried out in earthworms (*Eisenia foetida*). The test is divided into two parts:

1. *Acute paper contact test*: here the earthworms are exposed for 48 h to filter paper soaked in different concentrations of the test chemical. An LC_{50} is determined 48 h after commencement of exposure.

2. *Chronic artificial soil test*: the earthworms are exposed to varying concentrations of the test chemical in a standard artificial soil preparation (this allows adsorption and desorption to be standardised). An LC_{50} is determined after 7 or 14 days' exposure.

3.4.4.8 *Rapid Ecotoxicity Tests*

The major problem with the trophic level-based ecotoxicity testing strategy is that it takes a long time to test a new chemical and is very expensive (for a complete battery of tests it would cost £100 1000s). Rapid screening tests are therefore necessary to have a quick 'look see' at a new compound in order to decide whether to proceed with the more expensive studies. Clearly it is not possible to produce a definitive answer from such tests, but an indication of the potential environmental impact of the test chemical will be given.

Microtox® (manufactured by Microbics UK Ltd) is a screening test based upon the fluorescent marine bacterium *Photobacterium phosphoreum*. If the organism fluoresces (i.e. produces light by cellular biochemical reactions) it is alive and well; fluorescence decays when it is adversely affected by chemical challenge. This important property of the organism is utilised to good effect in assessing the potential environmental toxicity of a new chemical. In addition, the test is useful as it allows combinations of chemicals to be studied and the possibility of synergy to be investigated.

There is much controversy over the use of Microtox® and other similar tests; clearly they can only be used as general screens and must not replace conventional ecotoxicity testing procedures. Microtox®, however, has been shown to reflect well data on the toxicity of specific chemicals in higher test systems; indeed several countries (e.g. Canada) accept Microtox® data as part of regulatory packages used for approval purposes.

Microtox® tests are simple and quick to carry out. The test chemical is serially diluted in up to 15 tubes which fit into a purpose-built machine (see Figure 3.8), a

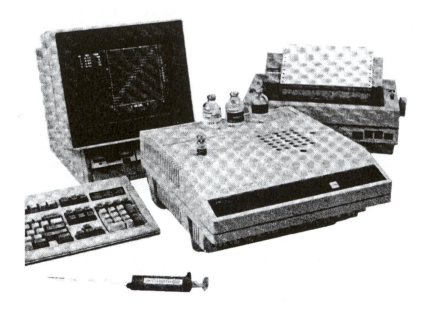

Figure 3.8 The Microtox® set-up showing the Microtox machine with a series of wells on top into which the tubes containing serially diluted test chemical are placed. A typical dose-response graph can be seen on the VDU. Reproduced by kind permission of Microbics UK Ltd, Hertfordshire, UK.

suspension of *P. phosphoreum* is added to each tube and the tube is then incubated at 15°C (the bacterium is from temperate seas and requires cool incubation conditions). After a short (5 or 15 min tests are usually carried out) incubation period the bacteria's phosphorescence is automatically measured by the Microtox® machine and plotted against the concentration of the test chemical (see Figure 3.9) and the EC_{50} is calculated from the data.

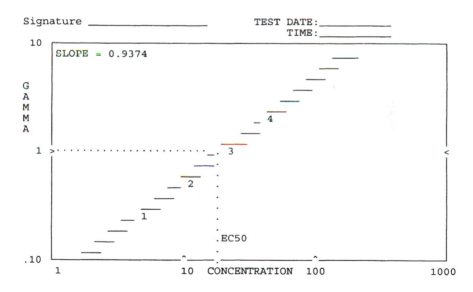

```
                        MICROTOX DATA REPORT
                            Basic Test
FILE: PHENOL.K5
PHENOL QA CONTROL

Test Time:   5 minutes                      Osmotic Adjustment:none

        NUMBER          IO/IT              CONC.        CR/GAMMA
        -------      ------------       -----------     ---------
        Control     95.23/ 94.12           0.0         0.9883 #

          1         95.34/ 72.07           5.0000      0.307 #
          2         96.11/ 59.14          10.0000      0.606 #
          3         89.88/ 41.13          20.0000      1.160 #
          4         85.98/ 26.89          40.0000      2.160 #

    CR = Control Ratio         CORRECTION FACTOR = 0.9883
    # Used for calculations

EC50    17.3254 mg/l   (95% CONFIDENCE RANGE:   16.5174 TO    18.1730)


Signature _____     TEST DATE:_____
                                        TIME:_____
```

ESTIMATING EQUATION: LOG C = 1.0664 x LOG Γ +1.2387
95% CONFIDENCE FACTOR: 1.04892   FOR EC50
COEFFICIENT OF DETERMINATION: R^2 =   0.99962

Figure 3.9 Typical Microtox® printout showing the phosphorescence dose-response curve for phenol. The $EC_{50} = 17.3\,mg\,dm^{-3}$.

3.5   Environmental Impact Assessment

It is far beyond the scope of this book to cover environmental impact assessment in detail. It is covered here in the context of how the ecotoxicity test results are used to assess the impact of a particular potential environmental pollutant upon the environment.

We will use a real example of a veterinary drug to illustrate the principle of environmental impact assessment (the drug has been given a false name); this calculation was used as part of the drug's Approval for Marketing Application. The purpose of the calculation is to determine if the test chemical (in this case the veterinary medicine Atonin) might reach concentrations in a particular ecosystem (in this case a pond) which might deleteriously affect members of the ecosystem.

Data

Drug:	Atonin
Use:	In-feed in pigs
Concentration in feed	$100 \, \text{g ton}^{-1}$
Potential routes of environmental contamination	
1. Spillage of feed	
2. In the pigs' excreta	
Number of pigs on typical farm	1000
Feed given $\text{day}^{-1} \text{pig}^{-1}$	2 kg
EC_{50} (*Daphnia*)	$1.2 \, \text{mg dm}^{-3}$
EC_{50} (trout)	$0.5 \, \text{mg dm}^{-3}$
NOEL (mallard)	$0.75 \, \text{mg kg}^{-1}$
Volume of typical pond	$10^4 \, \text{dm}^3$

Calculation

Total mass of Atonin given to pigs per day

$$0.002 \text{ (i.e. 2 kg in tonnes)} \times 1000 \times 100 \text{ g}$$
$$= 200 \text{ g}$$

Assuming the worst case, all would enter the pond

Concentration in pond
$$= \frac{200g}{10^4 dm^3}$$

$$= 20 \, \text{mg dm}^{-3}$$

Therefore in this example the worst case pollution incident would result in a concentration of Atonin in the pond which greatly exceeds the EC_{50}s and NOEL. It is very unlikely that approval for marketing would be granted.

Further Reading

Erickson, P.A., 1994, *Environmental Impact Assessment*, San Diego: Academic Press.

Karcher, W. and Devillers, J., 1990, *Practical Applications of Quantitative Structure Activity Relationships in Environmental Chemistry and Toxicology*, Chemical and Environmental Science Series, London: Kluwer Academic Publishers.

Organisation for Economic Cooperation and Development, 1993, *Guidelines for Testing of Chemicals*, OECD, Paris.

4

Environmental Monitoring

This chapter discusses the importance of monitoring the environment for pollutants using analytical chemistry. Food chain contamination and food analysis is discussed. The main body of the chapter is devoted to describing the analytical methods used for monitoring, including:

- Extraction procedures
- Gas liquid chromatography
- High performance liquid chromatography
- Mass spectrometry
- Nuclear magnetic resonance spectroscopy
- Infrared spectroscopy
- Atomic absorption spectrophotometry
- Immunoassays

Monitoring environmental contaminants involves analytical chemistry. Entire books have been written on this subject and some scientists have spent their whole careers studying single analytical techniques. For this reason this chapter can only be a very cursory overview of the subject. It is intended only to introduce the bones of the subject and to illustrate the importance of a knowledge of analytical chemistry to the environmental toxicologist.

4.1 Why Monitor Environmental Contaminants?

In order to assess the potential harm that environmental pollutants pose we need to know the environmental concentrations of the chemicals concerned. Way back in the mid-fifteenth century Paracelsus, an influential German scientist (his real name was Thophrastus von Hohenheim but, as did many of the scientists of the time, he assumed an intellectual Latin name), realised that quantities or concentrations (i.e. doses) of chemicals were important in determining their toxicity (see Section 2.3).

It is very well established that we need to know how much of a chemical is present before we can assess its likely deleterious effects upon organisms with which it might come into contact.

4.1.1 *Nitrate*

Nitrate is a very important nutrient which must be present in soil and water ecosystems to facilitate the growth of primary producers (which in turn support the rest of the ecosystem); its concentration in a healthy aquatic ecosystem would be $5 \, \mu g \, dm^{-3}$. If, however, the NO_3^- concentration reaches $50 \, \mu g \, dm^{-3}$ in a pond, the pond becomes eutrophic and the NO_3^- poisons the system by supporting the overgrowth of algae or other plants. Clearly at $5 \, \mu g \, dm^{-3}$ the NO_3^- is beneficial and at $50 \, \mu g \, dm^{-3}$ it is deleterious. Monitoring NO_3^- concentrations in water courses gives an idea of the extent of pollution from agricultural fertiliser run-off and might be used to predict the onset of a eutrophic state. There are many more examples (phosphate for instance) of nutrients which damage the environment at high concentrations.

4.1.2 *Organic Contaminants*

At the other end of the spectrum from nitrate there are ostensibly toxic chemicals (e.g. phenols from disinfectants) which have no measurable effects on ecosystems providing their concentration in the ecosystem does not exceed a well-defined level. Monitoring helps us to determine how the use of such chemicals causes their concentrations to rise and when we should attempt to restrict (or even ban) their use for fear of harm to the environment resulting from their increased concentrations. This use of monitoring is very important because it allows important industrial, household or agricultural chemicals to be used with confidence. It does, however, admit that we are prepared to live with such chemicals contaminating our environment: we have to hope that our initial assessment of the concentration at which such contaminants become toxic was correct. There are notable examples where scientists and regulators have been wrong.

4.1.3 *Contaminants in the Human Diet*

The acceptable concentration of a chemical in the environment is very loosely defined indeed because, for the most part, we do not know enough about environmental toxicology to predict at what concentration something might pose an unacceptable risk. The situation with food is quite different. As people are involved (and we have a vested interest in their survival) extensive studies in animals have enabled an idea of the concentrations of chemicals in food that are acceptable (on safety grounds). These concentrations are based upon the amount that can be ingested each day (for an entire lifetime) in a normally constituted diet which would result in no pharmacological or toxicological effects (the acceptable daily intake; ADI). This value plus a huge safety factor (usually 10^3) is used to calculate the maximum amount of a chemical that is allowable in a particular food item (the maximum residue level; MRL). The MRL gives the analyst a benchmark to which his analytical findings can be compared. If a result is above the MRL then in theory the food should not be eaten.

$$\text{ADI mg} = \frac{\text{NOEL mg kg}^{-1} \text{body weight}}{\text{SF 60 kg}^{1}}$$

NOEL   = No observable effect level in a sensitive test animal species (e.g. rat). This is the highest dose at which no biochemical or pharmacological effects are seen. It has replaced the old LD_{50} test

SF   = Safety factor (usually 10^3).

$$\text{MRL mg kg}^{-1} = \frac{\text{ADI mg}}{\text{Weight of food consumed kg}^{2}}$$

[1] The regulator's average human weight.
[2] This weight is derived from dietary survey data which allow the calculation of average daily consumption for specific foods, e.g. liver = 100 g

MRLs therefore give us a good safety standard which incorporates a huge safety factor. To confuse things, they apply only to chemicals which might find their way directly into our food (e.g. veterinary medicines given to food-producing animals which form residues in meat). Accidental contaminants (e.g. pesticides) are assigned MRLs, but in a very different way. The pesticide MRL is defined as the concentration of a particular pesticide which is present in a defined food plant following use of the pesticide in accordance with good agricultural practice (GAP). Whether or not we support the way in which pesticide MRLs are arrived at does not detract from the fact that it gives the analyst a benchmark for his analysis (see Table 4.1) and the regulators a value above which a violation has occurred and action should be taken.

Table 4.1 Some MRLs for commonly used pesticides in food

Food	Pesticide	MRL (mg kg^{-1})
Breakfast cereals	Pirimiphos methyl	5
Wheat	Malathion	8
Bananas	Imazalil	2
Goat's cheese	γ-Hexachlorocyclohexane	0.2
Strawberries	Carbendazim	5
Plums	Pirimicarb	0.5
Grapefruit	Metalaxyl	5

4.2 Methods of Monitoring

Before analysis of environmental samples can be started there are three important considerations:

1. What to sample: water, soil, plants, etc.
2. How many samples to take: statistical considerations
3. When to sample: seasonal variations.

4.2.1 *What to Sample*

The obvious thing to do is to take samples from the most likely facet of the environment to have been contaminated. This might be appropriate for studies to investigate, for example, whether a river has been polluted by a pesticide used on an adjacent field. Such studies are carried out and give an answer to a rather limited question. In this case water samples would be taken from the river and analysed. Here the extent of contamination of the environment as a whole following the use of a particular chemical is sought. Clearly, a broad sampling regimen to encompass as many ecosystems as possible is necessary. For example, if we were striving to assess the impact of lead from car exhaust fumes upon the environment a very broad sampling regimen would be necessary (see Table 4.2).

Sampling is extremely important, it is crucial to take samples from the right place to reflect a particular facet of the environment.

Table 4.2 Examples of samples necessary to assess concentrations (and therefore impact) of lead from car exhaust fumes

Sample	Purpose
Air near to major road	To determine whether lead is present in air near to 'car activity'
Air 500 m from major road	To determine whether lead levels in air travel from source
Soil from roadside	To determine sedimentation of lead from car exhaust
Soil 500 m from roadside	To determine whether sedimentation is related to air concentrations of lead
Vegetation from roadside	(i) To determine vegetation surface concentration of lead (ii) To determine whether lead is absorbed by plants
Water from rivers many kilometres from roadside	Determine 'global' lead contamination of waterways

4.2.2 *How the Sample is Taken*

How the sample is taken is of equal importance. If you are looking at lead contamination at the roadside, higher lead concentrations will presumably be achieved if surface soil is sampled. It might, however, be interesting (or even important) to investigate subsurface soil. Soil augers (see Figure 4.1) are used to take deeper soil samples.

If water is being sampled similar questions of depth are important, but additionally where in the waterway or pond the sample is taken is also important; should the sample be taken from near the bank or in the middle of the waterway? There is often no correct answer to this question; it is often necessary to sample several areas in order to find out if there are differences.

Figure 4.1 Soil profile showing the different layers that are likely to have different concentrations of environmental contaminants depending upon their elution properties through the soil. Reproduced from *Soils and Environment* by S. Ellis and A. Mellor, 1995, Routledge, by kind permission of the publishers.

4.2.3 *Number of Samples*

The statistics of sampling is important. Results will vary between individual samples and it is important to assess the magnitude of the variability (precision). If only one sample of soil in an environmental contamination study were taken and (for example) lead was found to be present at $200 \, \mu g \, g^{-1}$ the analyst would almost certainly be very worried, because from this single sample the only (rather pessimistic) conclusion that can be drawn is that this concentration occurs throughout the area. If, however, three samples were taken and the results were 10, 200 and 15 $\mu g\,g^{-1}$ this suggests that only part of the sampling area has been more heavily contaminated than the surrounding region and the analyst would perhaps then be less concerned. If, however, 20 samples were taken and the results were as in Table 4.3 a quite different conclusion would be drawn. In this case the experiment involved 20 samples. Samples 5 and 12 were heavily contaminated, sample 18 was less contaminated but still at least twice as contaminated as what appears to be the background lead concentration (i.e. 6-17 $\mu g\,g^{-1}$). That one region of the sampling site had been contaminated is still a possibility. If samples 5, 12 and 18 (see Table 4.3) were taken from neighbouring positions the contamination hypothesis gains credibility. Sampling position and extent *are* important in generating a scientifically meaningful result.

Some elegant studies have been carried out in the USA which clearly show different concentrations of a pollutant at varying distances from a point source (see Figure 4.2). The example shown in Figure 4.2 is for H^+ (i.e. acidification), the pH (i.e. $-\log[H^+]$) is lower (i.e. $[H^+]$ is greater) near to the industrial area from where the 'acid' gases (e.g. SO_2) are being released into the environment.

Table 4.3 Hypothetical analytical results for lead in soil samples, showing variability and mean values

Sample	Lead ($\mu g\,g^{-1}$)
1	10
2	10
3	15
4	12
5	200
6	10
7	15
8	17
9	10
10	8
11	8
12	100
13	10
14	12
15	15
16	13
17	6
18	40
19	13
20	12
Mean ± SD	

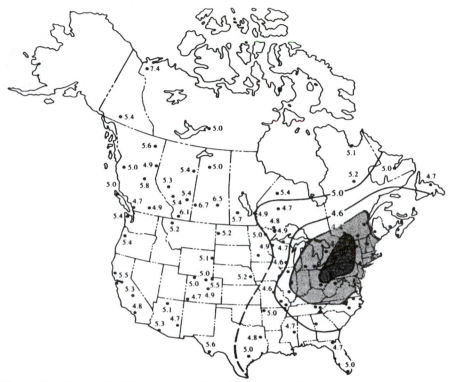

Figure 4.2 Map of the USA showing soil pH decreasing with decreasing distance from an industrial region. This is a good example of a point source of pollution. Reprinted from B. Freedman, 1989, *Environmental Ecology*, Academic Press Ltd, by kind permission of the publishers.

4.3 The Meaning of Analytical Results

It is tempting to regard analytical results as absolute. If an analytical chemist reports a DDT concentration in a soil sample as $0.01 \, \mu g \, g^{-1}$, what does this mean? Three parameters need to be known before we can really understand the result, namely:

1. *Limit of detection*: the minimum concentration of the analyte that can *reliably* be found in the particular sample type.

2. *Analytical variability*: if the analyst had analysed the same sample 10 times, how much would the result vary?

3. *Stability*: is the analyte stable in the analytical matrix?

If, for example, the limit of detection in the analytical procedure for DDT had been $0.02 \, \mu g \, g^{-1}$, then the reported result (i.e. $0.01 \, \mu g \, g^{-1}$) is meaningless because it is

Table 4.4 Results for thiabendazole in potatoes taken from the Working Party on Pesticide Residues Report (MAFF, UK, 1994) showing the importance of analytical parameters

Result (mg kg^{-1})	Number of samples in range
< 0.1 (not detected)	29
0.1	1
0.5	1
1.0	1

[a] Limit of determination = 0.1 mg kg^{-1}
[b] MRL = 5.0 mg kg^{-1}
[c] Not detected might mean anything between 0 and 0.1 mg kg^{-1}
[d] The analytical method used (high performance liquid chromatography) is well suited to determining thiabendazole in potatoes to an MRL of 5 mg kg^{-1}.

below the concentration that can be determined reliably. If the analytical variability were ±50 per cent, then the result could be anywhere within the range 0.005–0.015 $\mu g\,g^{-1}$ and if the analyte degrades when extracted from the matrix (e.g. soil) with a $t_{1/2}$ of 3 min the result depends upon how long it took the analyst to complete the analysis. If it took 3 min the result would be twice that if the analysis took 6 min to complete. All of these variables must be taken into account when using the results of environmental analysis (see Table 4.4).

4.4   Analytical Techniques

In order to analyse a sample, the contaminant (analyte) must first be extracted from the sample (matrix) so that it can be concentrated and subjected to an appropriate analytical method (see later in this chapter).

4.4.1   *Extraction Procedures*

The most straightforward sample to analyse is water. Most organic pollutants tend to be hydrophobic and so have a very limited solubility in water. It is therefore a relatively easy task to extract the contaminant from a water sample into an organic solvent (e.g. hexane) which is insoluble in water. This is carried out by shaking the sample of water with hexane (or some other organic solvent, e.g. chloroform, diethylether, dichloromethane, etc.) in a separating funnel, separating the aqueous from the organic phase, re-extracting the aqueous with solvent, repeating and combining the organic phases. The solvent can then be evaporated to concentrate the analyte (care must be taken; it is possible that the analyte is volatile too). Generally the solvent is completely evaporated to leave a residue which is

(a)

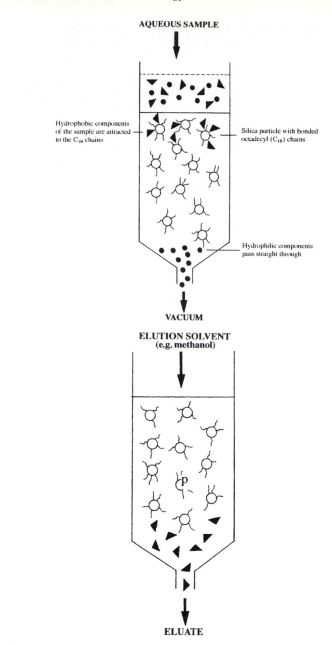

Figure 4.3 Schematic representation of the function of a solid-phase clean-up column. (a) The aqueous sample containing hydrophobic (▲) and hydrophilic (●) components is loaded onto the column and pulled through using a vacuum pump. The hydrophilic molecules pass straight through. The hydrophilic molecules are attracted to the bonded octadecyl chains and are eluted with an appropriate solvent (e.g. methanol). (b) Disposable column showing the small amount of packing material and the relatively large volume to accommodate the sample.

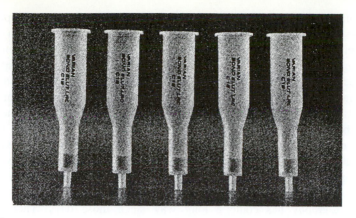

Figure 4.3b

dissolved in another solvent suitable for direct application to the analytical method of choice.

Soil, herbage and other environmental samples present a greater problem. The analyte may be adsorbed onto the surface of the matrix particles. For example, clay is an important component of soil and river silts, it has a net negative charge and can quite strongly bind positively charged analytes. It may therefore be necessary to use more drastic techniques than simply extracting with solvents. For instance, acids might be necessary to displace adsorbed analytes. The simplest extraction procedure is very similar to that described for water and involves shaking the sample with an organic solvent. Often, however, it is necessary to dry the sample in an oven before applying solvent extraction techniques. This is because the adsorbed environmental contaminant might be protected by a thin film of water adhering to the sample's surface and so not partition efficiently into the organic solvent.

4.4.1.1 *Solid-phase Clean-Up Columns*

Another way of concentrating the analyte is to pass a crude extract of the sample (or, for example, even a sample of pond water) down a specially designed chromatography column which adsorbs the analyte. The column can then be washed by passing a buffer or solvent down it to remove any interfering chemicals and then the analyte eluted with a solvent (see Figure 4.3). These columns can now be bought pre-packed and are disposable, which makes large-scale extractions a much simpler task.

4.5 Chromatography

Having extracted the analyte from the environmental sample and purified the extract to remove as many interfering molecules as possible, the sample is usually subjected to chromatography.

Chromatography is a science in its own right. Once it only meant separation of molecules using absorbent paper. The sample was applied at the bottom of a large (often 1m × 1m) sheet of chromatography paper and the whole developed in a chromatography tank with a small amount of solvent mixture (designed specifically for the separation to be achieved). The solvent moved up the absorbent paper differentially dissolving components of the mixture to be separated. Separation was dependent upon there being different partition coefficients of the components of the mixture between the paper and the solvent. Those that 'preferred' the paper migrated more slowly whereas those that were very soluble in the solvent mixture moved at or near the solvent front. This is the traditional concept of chromatography that most school children see illustrated by separating black or blue ink on blotting paper to reveal their brightly coloured components.

Chromatography has come a long way since those days. There are now sophisticated 'high-tech' chromatographic techniques, several of which are commonly used in environmental analysis.

4.5.1   Gas-Liquid Chromatography (GLC)

Here the paper (solid support) of the chromatography column is replaced by an inert support (the stationary phase) packed into a very long (usually 1–3 m) narrow metal or glass tube. A mixture of gases is passed over this and the components of the sample mixture are separated on the basis of their partition between the stationary phase and the gas. The whole system is (generally) maintained at high temperature (of the order of 100–200°C) to facilitate vaporisation of the sample. It is called *gas-liquid* chromatography because the partition is between the gas which is pumped through the column and a stationary oily liquid which is adsorbed onto the surface of the particulate stationary phase.

The main problem is that the analyte has to be volatile. As most analytes are not volatile it is necessary to derivatise them to increase their volatility and make them amenable to GLC analysis. There are many derivatisation procedures used, but just one commonly used example in pesticide residues analysis will be given here.

4.5.1.1   GLC Analysis of Diazinon

Diazinon (see Figure 4.4) is an organophosphorus (OP) insecticide which is used as both an agricultural pesticide and an ectoparasiticide in veterinary medicine (e.g. as a component of sheep dip to kill lice and the sheep scab mite). Residues of Diazinon are often measured by GLC. Diazinon is separated from other OPs by GLC (see Figure 4.5).

4.5.1.2   GLC Detection Methods

There are many GLC detection techniques and only the most commonly used will be described here. The difficulty in applying detection techniques to GLC relates to

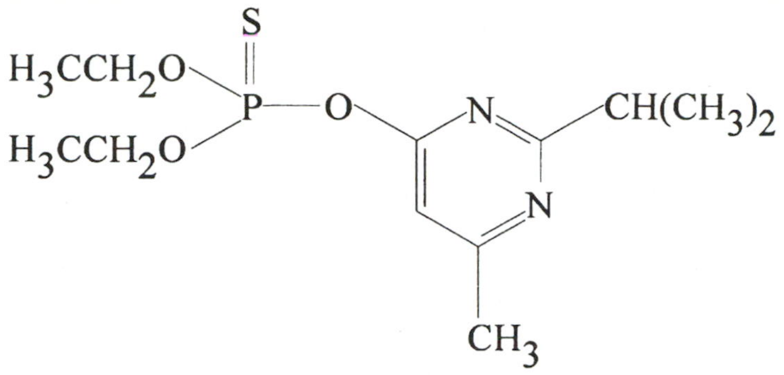

Figure 4.4 Molecular structure of the sheep dip OP, Diazinon.

the problem that the effluent from the column is in gas form and sometimes (because of the high temperatures used to maintain the gaseous state) the molecular structures of the chemicals being analysed change during the course of the analysis.

Flame Ionisation Detection (FID) The gaseous effluent from the GLC column is passed through a hydrogen flame. Molecules in the effluent (e.g. environmental contaminants) are ionised by this process and are attracted to a detector which responds to their charge. An electrical impulse is produced which is recorded. Its position in time corresponds to the V_R (*retention volume*) or T_R (*retention time*) of the particular component of the sample mixture and its magnitude is proportional to the concentration of the molecule.

Electron Capture Detection This is often termed ECD which might be confusing because electrochemical detection has the same abbreviation. ECD involves ionisation of the GLC carrier gas (e.g. N_2) by bombarding it with β-particles from a radioactive source, forming positive ions. The electrons displaced by ionisation of the carrier gas are attracted to a positive electrode and cause a current to flow. Molecules present in the carrier gas (i.e. the molecules being detected) 'steal' the electrons thus reducing the current flowing from the detector and are therefore measured as negative peaks. The electronics of the recording system reverse this to give conventional positive peaks.

Nitrogen Phosphorus Detection (NPD) As its name suggests the NPD is specific to nitrogen and/or phosphorus atoms in organic molecules. It is a very useful detection method for many environmental contaminants because nitrogen is a particularly common constituent of such molecules. An excellent example of a broad group of environmental contaminants which contain both nitrogen and phosphorus is the OPs.

The detector has a rubidium chloride bead which is heated to white heat. The effluent from the column is passed over the bead, nitrogen and phosphorus interact with the hot rubidium chloride bead and the current generated is detected.

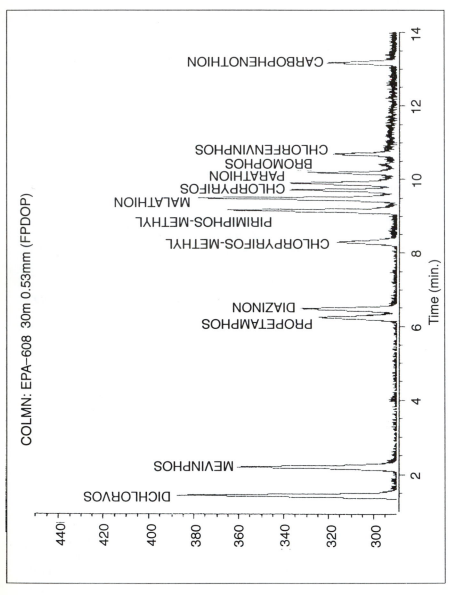

Figure 4.5 GLC trace showing the separation of Diazinon from other OPs. Data kindly provided by Richard Parker, Central Veterinary Laboratory, Weybridge, UK.

4.5.2 *High Performance Liquid Chromatography (HPLC)*

This is a more recent introduction (in the early 1970s) and might be thought of as a variant of GLC where the mobile phase is a liquid instead of a gas and the stationary phase is chemically bonded to the support.

The HPLC support is generally fine particle (3–5 μm diameter) silica with various chemical groups attached to it. The nature of the chemical group is varied according to the type of analytes to be separated. A commonly used group is octadecyl silane (ODS or C_{18}) which is attached to the hydroxyl groups of the silica support. This makes a very hydrophobic column and this is termed reverse-phase HPLC. Hydrophobic molecules partition into the immobilised liquid stationary phase (see Figure 4.6) and are retained longer than hydrophilic molecules by the column. Other stationary phases include -CN, -CH_2CH_3 and -$CH_2CH_2NH_2$. Unmodified silica columns are also commonly used; this is termed normal-phase HPLC.

As the stationary phase particle size is very small, the mobile phase has to be pumped at high pressure (about 1500 psi) to force it through the column. It is for this reason that the uninformed think that the P in HPLC stands for pressure. Performance refers to the column's incredible resolving power. The enormous surface area of the column readily separates very similar molecules. For example, phenol, uracil, acetophenone, methylbenzoate and toluene are well resolved on a C_{18} column with a methanol–water (10:90 v/v) mobile phase. The order of elution from the column is related to the polarity of the molecule. Separation is on the basis of partition between the stationary (hydrophobic) phase and the mobile (hydrophilic) phase. For reverse-phase HPLC with a relatively polar solvent, the more hydrophobic the molecule the longer it will remain on the column (i.e. its T_R, or V_R, will be greater). The order of elution therefore reflects the order of polarity. For the example of separation given above, the polarity order and elution order is shown in Figure 4.7.

The V_R is characteristic of a molecule (although it is possible that two molecules might have the same V_R in the same HPLC system) and so is often used in conjunction with standards (i.e. co-elution or co-chromatography) to suggest the identity of a peak obtained from (for example) an environmental sample.

This discussion of separation of substances by HPLC relies upon detection of the molecules. There are many ways that detection is achieved, but only the two most important in environmental monitoring will be considered here.

4.5.3 *HPLC Detection Methods*

4.5.3.1 *Light Absorption*

Molecules that are coloured can be detected by passing the HPLC column effluent through a flow-through spectrophotometer set at the appropriate absorption wavelength. This is not commonly used because detection sensitivity is often low and (more importantly) most environmental contaminants are colourless.

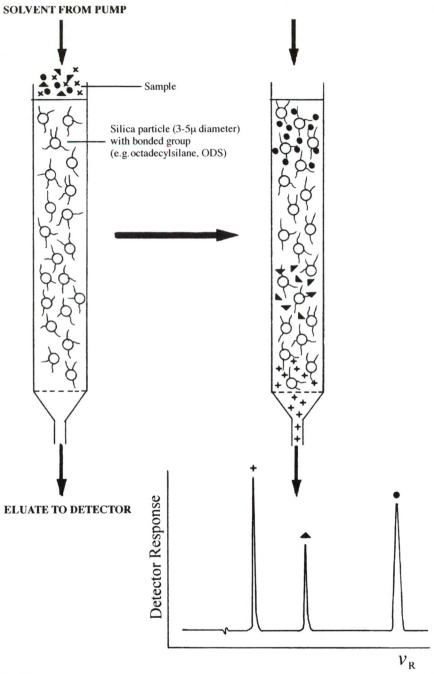

Figure 4.6   Schematic representation of the principles of reverse-phase HPLC. In this example the sample contains three different molecules in the following order of increasing polarity: ●▲✖.

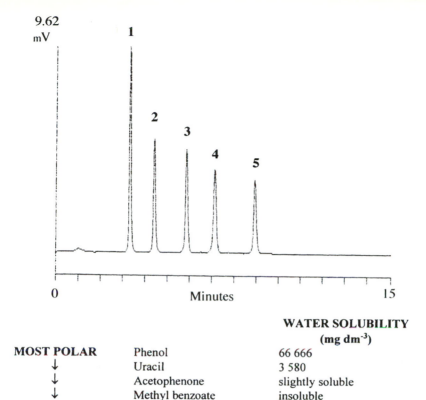

9.62
mV

0   Minutes   15

WATER SOLUBILITY		
		$(mg\ dm^{-3})$
MOST POLAR	Phenol	66 666
↓	Uracil	3 580
↓	Acetophenone	slightly soluble
↓	Methyl benzoate	insoluble
LEAST POLAR	Toluene	insoluble

Figure 4.7 Polarity order of five molecules showing their separation by reverse-phase HPLC which reflects* this order of polarity. HPLC system: Column, 30 cm C_{18} reverse-phase (Spherisorb ODS®); mobile phase, methanol-water (70:30 v/v); detection A_{254nm}. HPLC trace kindly provided by Jones Chromatography Ltd, Glamorgan, UK. 1: uracil; 2: phenol; 3: acetophenone; 4: methylbenzoate; 5: toluene.
* Phenol and uracil are apparently reversed; however, uracil is more polar than phenol despite phenol's apparent greater water solubility — phenol is peculiar in that water effectively dissolves in it so giving the impression of high water solubility.

Ultraviolet (UV, i.e. $\lambda < 340$ nm) absorption is the most common detection technique applied to HPLC in environmental analysis because most organic environmental contaminants absorb in the UV range. A similar set-up to that described for visible detection is used except that the spectrophotometer has a mercury vapour lamp to generate light at the UV end of the spectrum.

4.5.3.2 *Electrochemical Detection (ECD)*

This is an extremely sensitive detection technique and relies upon the molecule to be detected being either oxidisable or reducible. Oxidation ECD is by far the commonest mode and therefore we only deal with this here.

A potential difference (PD; usually between 100 mV and 1 V) is applied across the effluent from the HPLC column. This results in oxidation of specific chemical groups on the molecule (e.g. $-OH \rightarrow =O$). This oxidation process consumes electrons which are detected by an electrode. Phenols are eminently well suited to ECD. The PD at which oxidation occurs is quite specific to the molecule and so a degree of detection specificity can be introduced by carefully selecting the ECD PD; a general rule is that at higher PDs more molecules are detected.

4.5.3.3 Quantification

The analytical procedures discussed above are used to tentatively identify contaminants by using co-chromatography (as discussed above), but much more importantly are used to determine the amount of substance present in the sample and therefore to determine the environmental concentration of a particular pollutant. The HPLC detection techniques rely upon Beer–Lambert's law which in essence states that optical absorption is proportional to concentration. The areas under the HPLC peaks are dependent upon light absorption (optical density) and, in turn, are proportional to the concentration of the molecule which is responsible for the peak. Peak area is therefore proportional to the concentration of the molecule 'in' the peak. Using a calibration graph (i.e. peak area for analytical standards versus concentration or mass) it is easy to calculate the concentration of a particular pollutant in an environmental sample.

Knowing the amount of substance present in a sample is essential to allow the risk associated with exposure to the sample to be determined. HPLC is also very important here because the peak area is proportional to the amount of substance present (for very sharp and symmetrical peaks the peak height may be used). Most modern HPLC equipment has a built-in integrator which automatically gives peak areas for the principal peaks in a particular separation. A calibration graph can be constructed and the unknown substance concentration in an environmental sample can be calculated.

4.6 Identification of Environmental Contaminants

So far we have been able to suggest the identity of environmental contaminants by comparing their V_R values from different chromatographic systems to those of authentic standards. This is by no means a positive identification and of course is of no use if one has no idea of the identity of a peak on a chromatogram of an environmental sample. If this were the case (and it often is) techniques which give information about the molecule itself have to be used in order to build up a picture that might eventually identify the unknown substance.

There is a battery of spectroscopic techniques which can be used. Only the most important in environmental analysis will be discussed here.

The most useful spectroscopic techniques are those which can be interfaced with the chromatographic procedures described above. This is important because it allows the resolving capacity of the chromatography to be linked to the identification of the spectroscopy. This means that you can have an unknown peak on a chromatogram of an environmental sample and 'ask' the spectroscopic technique to tell you what the peak is.

4.6.1 Mass Spectrometry (MS)

This is by far the most commonly applied procedure to the identification of unknown environmental contaminants. Amazingly MS can now be interfaced with both GLC and HPLC. Why this is amazing is because both of the chromatographic systems operate at high positive pressure and the MS operates in a vacuum. Clearly there is complex interface engineering necessary.

The MS works by bombarding a vaporised molecule in a vacuum with either electrons (most commonly) or positive ions (e.g. a positive ion derived from *iso*butane, $(CH_3)_3C^+$) which results in fragmentation of the molecule (a bit like shooting a flower pot with a shot gun; see Figure 4.8). The molecular fragments are charged and their mass to charge ratio (m/z) determines their trajectory as they move towards a charged plate (the detector). The plate is curved and so the position at which they impact the plate is dependent upon their m/z. As all of the fragments have the same charge (i.e. 1) it is their mass which determines their position of collision with the detector. Each time an ion hits the detector a signal is generated which is converted to a printout that shows the molecular fragmentation pattern (see Figure 4.9). For small molecules ($< MW = 500$) the highest m/z peak is usually the molecular ion and is equivalent to the molecular weight of the compound.

An experienced mass spectrometrist can sit with a spectrum and a calculator and determine the molecular structure of an unknown compound by calculating the difference between the m/z values for adjacent peaks and so determine which chemical group had left the molecule when under fire by electrons in the mass spectrometer (e.g. if the difference is 17 it is likely that OH has been lost; i.e. O = 16, H = 1, 16 + 1 = 17). By piecing this evidence together an identification can be made.

The know-how of an experienced mass spectrometrist has been incorporated into sophisticated databases which are usually part of modern MS systems. Using such databases the MS itself 'suggests' an identity for the unknown molecule.

4.6.2 Nuclear Magnetic Resonance (NMR) Spectroscopy

To understand NMR at its simplest level you must consider atomic nuclei as tiny magnets. If they are placed in a magnetic field they will vibrate and eventually align.

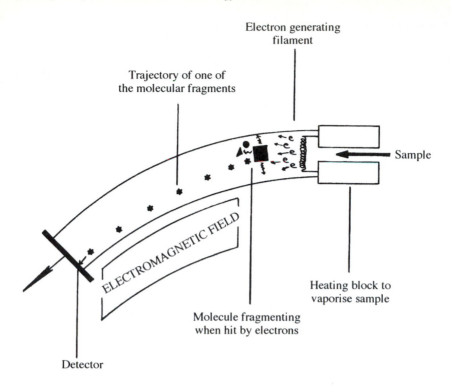

Figure 4.8 Schematic representation of the workings of a mass spectrometer.

The resonant frequencies for identical atomic nuclei in a constant external magnetic field vary with the chemical environment in which the particular nucleus is found. So, for example, a nucleus next to a hydroxyl group will resonate at a different frequency to the same nucleus adjacent to a methyl group. These differences are called chemical shifts and are important in determining the chemical environment of a particular nucleus in a molecule.

In the discussion so far we have been careful not to be specific about the identity of the atomic nucleus; in fact there are only a limited number of nuclei which are amenable to NMR spectroscopy because not all nuclei have magnetic properties. Nuclei of atoms with odd mass and atomic number have magnetic properties and therefore are useful to the NMR spectroscopist. The most important of these are ^{1}H, ^{19}F, ^{13}C and ^{31}P. In environmental monitoring ^{1}H (i.e. proton) NMR is by far the most common technique because protons are a very important component of all organic molecules.

So what exactly happens in the NMR spectrometer and how is the technique performed? The sample is dissolved in a solvent which has no ^{1}H in its molecular structure because such protons would interfere and make interpretation of the NMR spectrum very difficult. Generally deuterated (i.e. ^{2}H or D) solvents, such as

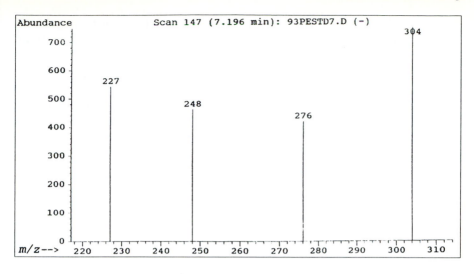

Figure 4.9 Electron impact fragmentation pattern for the organophosphorus pesticide Diazinon clearly shows the molecular ion at $m/z = 304$ which corresponds to the compound's molecular weight. The ions at $m/z = 276$ and 248 represent loss of the ethyl groups. Spectrum kindly provided by Richard Parker of the Central Veterinary Laboratory, Weybridge, UK.

deuterated chloroform ($CDCl_3$), are used. The material for analysis (probably isolated from an environmental sample by collecting fractions from an HPLC column) is dissolved in the deuterated solvent and placed in the NMR spectrometer where it is subjected to an enormous magnetic field (e.g. 4 Tesla; this is many times the earth's gravitational field). The sample tube is surrounded by an induction coil which generates a variable magnetic field at right angles to the constant magnetic field. The protons absorb energy from the field and align and in so doing generate a resonant frequency (RF) (see Figure 4.10).

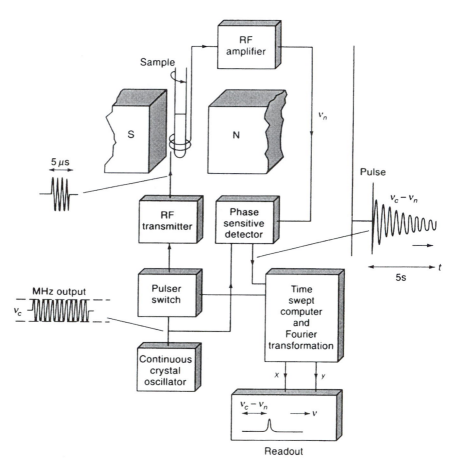

Figure 4.10   Schematic representation of an NMR spectrometer. The North (N) and South (S) poles of the magnet can be seen on either side of the sample tube. The sample is revolved to present an homogeneous picture to the detector. The magnetic field is pulsed with a particular frequency and the echoed signal is detected and amplified. The accrual and summation of numerous tiny signals (Fourier transform) allows low concentration samples to be studied. Reproduced from *Principles of Instrumental Analysis*, by D.A. Scoog and J.L. Leary, 1992, 4th edn, Harcourt Brace College Publishers, by kind permission of the publishers.

Table 4.5 NMR chemical shifts of protons in specific chemical groups (relative to the NMR standard tetramethyl silane)

Specific proton (underlined)	Chemical shift (p.p.m.)
H̲C-OR	3.4–4.0
RCOO-C̲H	3.7–4.1
H̲C-COOH	2.0–2.6
H̲C-C=O	2.0–2.7
RO̲H	1.0–5.5
R-N̲H$_2$	1.0–5.0
Ar-H̲	6.0–8.5
RC̲H$_3$	0.9

R: alkyl group; Ar: aromatic group.
Data kindly provided by Dr Peter Wearden, University of Central Lancashire.

The resultant NMR spectrum gives a great deal of information about molecular structure from which it is often possible to identify a molecule. NMR in conjunction with MS is very powerful indeed; MS determines the chemical groups present and NMR helps to pinpoint their arrangement in a particular molecular structure.

Interpretation of the NMR spectrum involves consideration of the type of spectral peaks. A single peak is given by a group of protons all influenced in exactly the same way by surrounding chemical groups (e.g. -CH$_3$) whereas a triplet would be produced by a group of protons differentially affected by their adjacent chemical groups. Specific groups in specific environments have specific chemical shifts. Tables are used to identify chemical groups in this way (see Table 4.5).

The area under the peaks is proportional to the number of protons in a particular category (see Figure 4.11) and therefore by integrating the peaks it is possible to determine the proportion of protons of different types in a molecular structure. This is very useful indeed in elucidating the molecular structure. Consider diethyl ether (ethoxyethane; CH$_3$CH$_2$OCH$_2$CH$_3$); there are 10 protons in total which would be in the ratio 6:4 on the NMR spectrum. It is clear from the spectrum of this molecule that the expected ratio is indeed seen (see Figure 4.12).

With the incredible developments in interfacing technology and computer software which enable suppression of the massive NMR signals from chromatographic solvents, it is now possible to combine NMR with HPLC. This is a very new development which is far from being routine, but the power of such a liaison will no doubt lead to rapid developments in this exciting field.

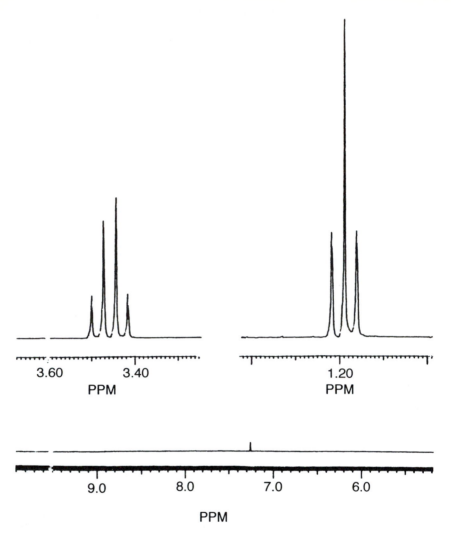

Figure 4.11 NMR spectrum for ethoxyethane ($CH_3CH_2OCH_2CH_3$): explanation of the identity of the spectral peaks. The quadruplet on the left represents the CH_2 protons and the triplet on the right represents the CH_3 protons. Spectrum kindly provided by Dr Peter Wearden, University of Central Lancashire.

4.6.3 *Infrared (IR) Spectroscopy*

This technique is far less commonly used in environmental monitoring, perhaps because of the relatively large amount of material needed. Nevertheless this can be a powerful technique in identifying environmental contaminants.

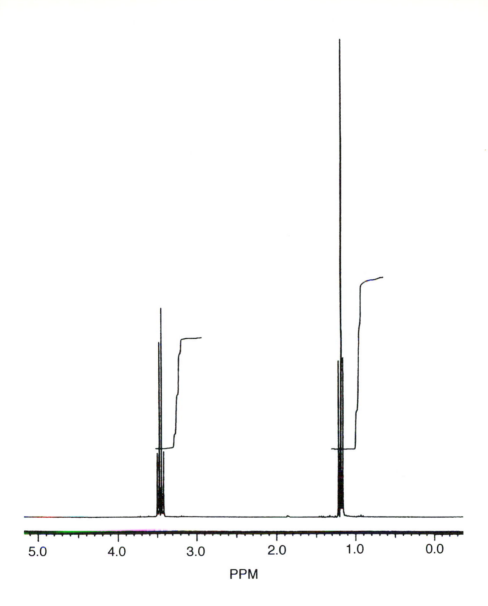

Figure 4.12 NMR spectrum of ethoxyethane (CH₃CH₂OCH₂CH₃) showing integration of the spectral peaks to produce the proton ratio 3:2 (i.e. 6 methyl protons and 4 methylene protons) and so helping considerably to elucidate the structure of the molecule. The methylene (C$\underline{H}_2$) protons are at 3.5 p.p.m. and the methyl (C$\underline{H}_3$) protons are at 1.2 p.p.m. The height of the integration line is proportional to the number of protons in the group.

Table 4.6 Infrared spectral characteristics of specific chemical bonds

Bond	Wave number (cm^{-1})
C-C	1200
C=C	1650
C≡C	2200
O-H	3600
C≡N	2200
N-H	3400
C-O	1100
C=O	1700
C-H	3000

These values are very useful in elucidating the structure of molecules; however, it is important to note that several different bonds have the same or very similar wave numbers (e.g. C≡C and C≡N). Data kindly provided by Dr Peter Wearden, University of Central Lancashire.

The technique involves isolation of pure contaminant (perhaps by HPLC) and then either smearing the substance in a pure hydrocarbon between potassium bromide discs (because hydrocarbons and potassium bromide do not interfere with the technique) or dissolving the substance in solvents (e.g. CCl_4) which do not interfere. The sample is then placed in the path of a beam of infrared radiation (i.e. $\lambda > 600$ nm). This causes the chemical bonds to stretch and vibrate and so absorb energy from the radiation causing a λ shift in the transmitted radiation. The result is a spectrum with bands in specific positions which are characteristic of the particular chemical group. Using tables the component chemical groups of a molecule can be determined (see Table 4.6) and so help in piecing together the identity of the unknown substance. *m*-Cresol has a characteristic IR spectrum which can be used as a fingerprint for the molecule (see Figure 4.13) and illustrates how facets of the molecule can be identified using IR peaks.

4.7   Inorganic Contaminants

So far we have only discussed techniques aimed at detecting and identifying organics (although MS can be applied to metals), but there are considerable concerns about contamination of the environment with metals from mining activity (e.g. arsenic as a waste product of the silver mining industry), lead from petrol and tin as tributyltin oxide which is used in antifouling paints in the marine industry. There are specific techniques available to facilitate the monitoring of metals and other inorganic ions (e.g. NO_3^-) in environmental samples.

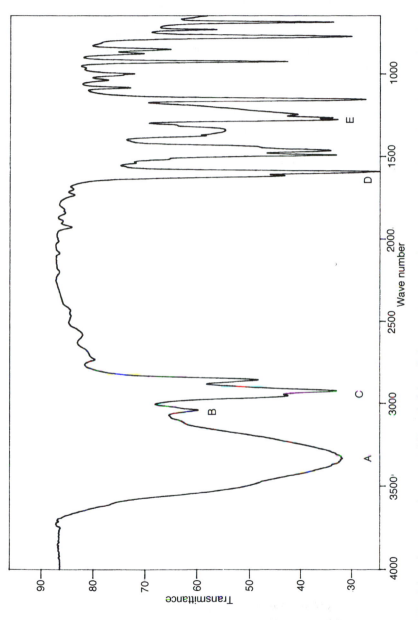

Figure 4.13 Infrared spectrum for *m*-cresol (3-methylphenol) showing the chemical groups responsible for the main spectral peaks. A: O-H; B: C-H (aromatic); C: C-H (aliphatic); D: C=C; E: C-O. Data kindly provided by Dr Peter Wearden, University of Central Lancashire.

The procedure most commonly applied to environmental samples to monitor inorganics involves initial isolation of the ions. Isolation often utilises ion exchange chromatography where charged columns (negative for metals — cation exchange, positive for anions — anion exchange) are used. The environmental sample (e.g. river water) is passed down the column, the ions are attracted to the opposite charge of the column packing material (adsorption) and the column is washed with a solvent (e.g. a buffer) to remove everything that has not adsorbed. The adsorbed ions are then displaced by ions which bind more strongly to the column (e.g. H^+ for metal ions) and thus the metals present in the water are isolated (see Figure 4.14).

Having separated the ions from other constituents in complex environmental samples using ion exchange chromatography the isolated substance, for example, metals can be identified and quantified by atomic absorption spectrophotometry (AAS) or the ion exchange column can be coupled directly to an ECD (e.g. Dionex®) and the eluted metals quantified directly. In this case their V_R values are compared with standards to facilitate identification of metals.

4.7.1 *Atomic Absorption Spectrophotometry (AAS)*

A solution of the metal is sprayed into a H_2/O_2 flame and the emission spectrum of the light is measured. Metals produce characteristic emissions when burned (e.g. copper gives a blue/green flame, potassium a purple flame and barium a green flame; this property of metals is used by fireworks manufacturers). The spectral properties identify the metal and the intensity of the light emitted is used to quantify the metal. This is a very useful technique which has developed to levels of great sophistication and allows the detection of very low concentrations ($\mu g\,dm^{-3}$ and below) of metals in solution.

This catalogue of analytical procedures only touches the surface of those used by the environmental chemist, but gives an idea of the level of sophistication involved in quantifying and verifying environmental contaminants. It represents the non-biological or pure analytical chemical techniques. Biology, however, still holds the key to analytical specificity and it is for this reason that immunoassays are also important in environmental monitoring.

4.8   Immunoassays

Antibodies are complex proteins produced by animals in response to foreign compounds. They are produced naturally in response to viral, bacterial or protozoal infections as a means of combating the infection. In order for a molecule to ellicit an antibody response it must be large (i.e. many 1000s of daltons). The analyst has overcome this problem by combining small molecules (haptens) with proteins

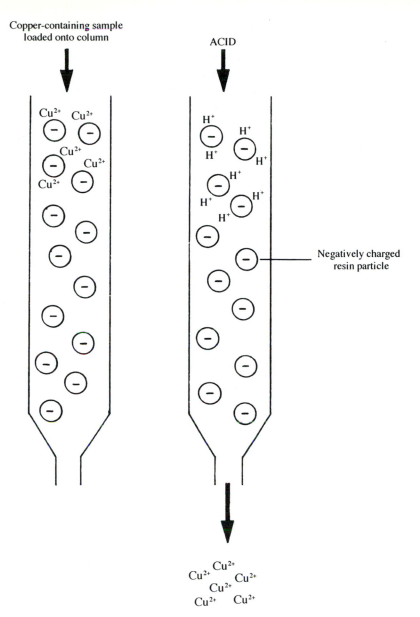

Figure 4.14 Schematic representation of the principle of ion exchange chromatography applied to the isolation of Cu^{2+} from a sample of river water. The sample containing Cu^{2+} is loaded onto the column. The Cu^{2+} is attracted to the negatively charged resin particles, other components of the sample pass through the column to waste. The column is washed and then eluted with acid. The H^+ displaces the Cu^{2+} which is eluted and collected.

(carriers) to produce a hapten–carrier complex, which when injected into an animal (usually a rabbit, sheep or goat) produces antibodies directed against the complex. A complex mixture of antibodies is produced among which it is very likely that there will be specific antibodies directed against the hapten. These antibodies are very useful indeed to the analyst in the detection of the hapten in environmental samples. The antibodies produced by this technique are termed polyclonal antibodies.

It is also possible to generate antibodies by molecular biological techniques in cultured cells. This is very much more specific and results in the production of a single antibody directed against a specific molecule. These are monoclonal antibodies.

The technique used to produce the antibodies is not really very important (in the context of our discussion) because the immunoassay procedure is the same irrespective of the source of the antibody.

There are several types of immunoassay which all depend upon an antibody directed against the molecule to be analysed. Such techniques are relatively uncommon in the environmental analyst's laboratory, but one or two are used to detect environmental contaminants at exceedingly low concentrations, for example dioxins and antibiotics such as penicillin.

4.8.1   Enzyme-linked Immunosorbant Assays (ELISA)

Here, in very simple terms, the antibody is linked to a high specific activity enzyme (e.g. horseradish peroxidase which catalyses the generation of reactive hydroperoxides from H_2O_2). This complex is added to the solution for analysis, the antibody reacts with the analyte (antigen) in solution and precipitates, and the precipitate is isolated and assayed for peroxidase activity. The peroxidase activity is proportional to the amount of antigen originally present in the solution. This technique can be made extremely sensitive by coupling numerous peroxidase molecules to each antibody, so amplifying the response. The enzyme assay is usually spectrophotometric making the assay relatively easy to carry out. Kits are often available for specific ELISA assays.

4.8.2   Radioimmunoassays (RIA)

The principle of RIA is very similar to ELISA; however, in RIA the antibody is made radioactive by iodination with ^{131}I of tyrosine residues in the antibody's protein molecule. The procedure is the same as for ELISA, but precipitated radioactivity is detected by γ-counting rather than enzyme assay. In this case the sensitivity is increased by iodinating more tyrosine residues.

The only problem (which is also its great advantage) with immunoassay is its amazing specificity. It clearly can be used only for routine monitoring of large numbers of samples because the cost of generating antibodies is great and therefore cannot be justified for one-off analyses.

4.9 Interpreting the Results of Environmental Analyses

Having analysed a series of samples from, for example, a river it is necessary to interpret the results in terms of the relevance of the concentration of the particular analyte to the well-being of the ecosystem. This process is perhaps best illustrated by a real example. In 1988 a major fire occurred in a wood treatment works in Surrey (UK). The treatment works was on the side of a tributary of the River Bourne which flows into the River Thames near to Weybridge. There is a water extraction point near the intersection of the rivers where water is taken for human consumption. The wood treatment works used γ-HCH and tributyltin oxide (TBTO) to protect wood against insect attack. The fire resulted in a large amount of the wood treatment fluid entering the tributary of the River Bourne, then the River Bourne itself and eventually the River Thames. Samples were taken from each of the rivers as soon as the spillage was notified. They were analysed by solvent extraction followed by GLC and MS for γ-HCH and analysis of tin for TBTO.

The effects upon the river's inhabitants are dependent upon the concentrations of γ-HCH and/or TBTO (see Table 4.7). Both contaminants rapidly disappeared from the aquatic phase of the ecosystem. Further analysis demonstrated that they had (at least in part) adsorbed onto the silt at the bottom of the rivers. This sequestration might well have been reversible and therefore it was decided to dredge the areas of the rivers with the highest silt levels of γ-HCH or TBTO and bury the contaminated silt in a landfill site.

A great concern at the time was the risk that extraction of water from the River Thames might pose to the population supplied with contaminated water. We will consider here only the situation relating to γ-HCH for which the ADI is 0.0005 mg kg^{-1} body weight. This means that only 0.03 mg can be ingested by an average person (i.e. weighing 60 kg) in 1 day. This is a very small amount. If we assume that an average person drinks 3 litres of water per day (this is an overestimate, but toxicologists work on worst case examples) the permissible concentration in water is:

$$\frac{0.03}{3}\text{mg dm}^{-3} = 0.01\text{mg dm}^{-3}$$

The actual concentration in the River Thames near to the abstraction point was well below this and therefore there was no problem. Despite this it was decided not to use the water for human consumption because there were alternative sources for the population in question.

Table 4.7 Data showing the concentrations of TBTO and lindane (γ-HCH) which kill the common ecotoxicity test species compared with the concentrations of the two pesticides in water from the River Bourne after a fire at a wood treatment works situated on one of the Bourne's tributaries

(a) Concentrations of TBTO and lindane which effect wildlife

Species	Concentration causing acute toxicity (LC_{50}, μg dm^{-3})	
	TBTO	Lindane
Rainbow trout (*Salmo gairdneri*)	6.9 (96-h exposure)	
Brown trout (*Salmo trutta*)		2 (96-h exposure)
Daphnia magna	1.67 (48-h exposure)	10 (48-h exposure)
Freshwater shrimp (*Gammarus pulex*)		30 (48-h exposure)
Water bug (*Notonectes* spp.)	30 (48-h exposure)	

(b) Pollution in the River Bourne

Day after incident	TBTO concentration (μg dm^{-3})	Lindane concentration (μg dm^{-3})
1	1750	300
2	200	
3	175	
7		150
30		60

Summary

The purpose of this chapter is to give an overview into the role of the analytical chemist in environmental toxicology. Environmental toxicologists must have a good knowledge of the analytical procedures used to generate data on levels of environmental toxicology which they will use as part of their risk assessments. This understanding is essential so that they are aware of the limitations of the analytical methods and therefore can draw conclusions about environmental impact with confidence.

Further Reading

Christian, G.D., 1994, *Analytical Chemistry*, New York: John Wiley .
Fifield, F.W. and Haines, P.J., 1995, *Environmental Analytical Chemistry*, London: Blackie Academic and Professional.

5

Human Toxicology

This chapter focuses on a single member of the biosphere, humans. It covers the assessment of the risk of environmental contaminants upon humans. Mammalian toxicity testing is discussed:

- Acute toxicity testing
- Subacute toxicity testing
- Chronic toxicity testing
- Reproductive toxicity testing
- Mutagenicity assays

The chapter closes with a discussion on the routes of exposure to environmental chemicals.

5.1 Humans as Members of an Ecosystem

If we consider the biosphere as the earth plus all of its inhabitants, humans must be regarded as an important part. The word important in this context has dual meaning. It refers to humans as animals, albeit manipulative animals able to modify their environment to suit themselves, and as the controllers or perhaps even owners of the environment. Whichever role we see ourselves in it is important that we are considered as a fundamental facet of the environment. On the other hand our environment might be regarded as the immediate ecosystem that we actually occupy, for example a man in a factory occupies a factory ecosystem. This ecosystem is more conventionally referred to as the workplace. People might be apparently the only living inhabitant of this ecosystem (except perhaps the factory cat), but almost certainly there are numerous other organisms occupying niches within this very specialised ecosystem. One thing is certain, humans as the controllers of the workplace are polluting it and putting themselves at risk of exposure to the deleterious chemicals that they use in their manufacturing processes. It is for this reason that we will look briefly at occupational exposure and the diseases associated with specific jobs.

In a more global sense people are exposed to chemicals in less direct ways.

Pesticides, medicines and the waste from both the home and industry eventually find their way into rivers, streams, the sea, land and air and so exert their effects upon humans via the food chain or by direct exposure. These issues form important aspects of legislation and Government Advisory Groups and Working Parties have been set up to monitor the chemicals that contaminate our world (e.g. the Working Party on Pesticide Residues which looks at pesticides in our food). We return to these later.

Before it is possible to draw meaningful conclusions from particular concentrations of environmental contaminants we must understand the toxicology of the contaminants and in particular the doses that are necessary to cause harm to (in this case) us.

5.2   Assessment of Risk to Humans

Assessment of risk to humans is no different, in principle, to the assessment of risk to any other species (see Chapter 3). There is, however, one major operational difference – it is usually not possible to study toxicity in the target species. Having said this though, accidental exposure and the use of medicines often provide useful information about the effects of specific chemicals upon humans *per se* (e.g. human exposure to 1-naphthylamine in the dye industry demonstrated unequivocally that it is a human bladder carcinogen). The ethical objections to studies of toxicity in people have not always existed. There are several periods in history (e.g. the Nazi period in Germany in the late 1930s and early 1940s) when objectionable studies in humans were carried out.

Studies in humans are rarely used now to assess toxicity. Animal studies are used and extrapolated to humans; this is an area of heated debate amongst toxicologists and regulators – which species best represents humans? The honest answer is, none. Risk assessment has to be based on studies in several unrelated species (e.g. rat and dog) followed by an expert assessment of the relevance of the results to humans.

Toxicity studies in animals are used to assess the potential harm of human medicines (which have direct effects on humans), veterinary medicines (to which people might be exposed via the food chain), pesticides (to which people might be exposed directly when using pesticides or walking through a treated field, or indirectly via the food chain) or industrial chemicals (to which people might be exposed directly in the workplace or indirectly by their contaminating water systems). Such toxicity studies assess hazard (i.e. the intrinsic toxicity of a chemical). Risk is determined by considering the likelihood (probability) of exposure to a particular hazard:

Risk = hazard × probability of exposure

Exposure to a chemical which has a high hazard (i.e. is very toxic, e.g. potassium cyanide, KCN) might have a very low risk if the person is exposed to a very low concentration of the chemical. It is quite safe (i.e. low risk) to drink

0.00000001 M KCN, whereas the risk associated with drinking 0.1 M KCN is enormous (in fact you would definitely die within minutes).

Hazard is assessed by a battery of animal tests which are discussed in outline below.

5.2.1 *Acute Toxicity Testing*

Acute toxicity refers to the immediate (acute comes from the Latin *acuere* to sharpen – coming sharply to a crisis) effects of a chemical following a single dose. Continuing with the KCN example, its acute toxicity results in death. Acute toxicity studies look at an endpoint and do not investigate the mechanism by which the eventuality comes about (e.g. KCN inhibits the terminal oxidase system and so prevents cellular energy metabolism). The most important outcome of the acute toxicity test is the dose of the chemical necessary to cause a specified acute effect. In the old days (before we became concerned about animal welfare) the most commonly observed acute effect was death. The outcome of an acute toxicity test was the LD_{50} (i.e. the dose of test chemical necessary to kill 50% of the test animal population). Thankfully our concern for the animals used in our tests has become of paramount importance in designing the studies and now the LD_{50} test is frowned upon (and not required by most regulatory authorities). Now instead of using death as the endpoint of the acute toxicity test some other parameter which indicates a physiological, biochemical or pharmacological adverse effect is used. For example, if liver damage is expected to be the route by which a test chemical exerts its toxicity, a change in the activity of a liver enzyme might be used to show when the test animal has begun to show signs of toxicity. A commonly used enzyme marker for the liver is glutamate pyruvate aminotransferase (GPT) (see Figure 5.1). In the normal healthy state the enzyme is within the liver cells (hepatocytes); however, when liver damage occurs the hepatocytes lyse and release GPT into the circulatory system. Measuring serum GPT (SGPT) activity therefore gives a good indication of liver damage. The dose of test chemical which does not result in an increase in SGPT (in this example, many other markers of toxicity might be used) is termed the no observable effect level (NOEL; see Figure 5.2). The NOEL has now replaced the LD_{50}.

Throughout our discussion of acute toxicity testing we have referred to the dose of the test chemical. The result of the acute toxicity test (i.e. the NOEL) is very closely related to the route of administration of the test chemical. For example, a chemical administered via the intravenous (iv) route is very likely to have a very much lower NOEL than the same chemical administered subcutaneously (sc). This is because toxicity is related to the concentration of the test chemical at its site of action (in our example the hepatocyte). Intravenous administration results in rapid delivery to most cells in the body, whereas the subcutaneous route would first require absorption, diffusion to capillaries, transport to the main circulatory system and then delivery to the target cells. This all takes time and results in a lower concentration of the test chemical at the target site (see Figure 5.3). When referring

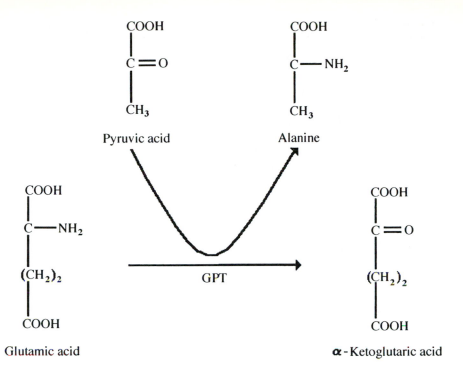

Figure 5.1 Biochemistry of the glutamate pyruvate aminotransferase (GPT) reaction

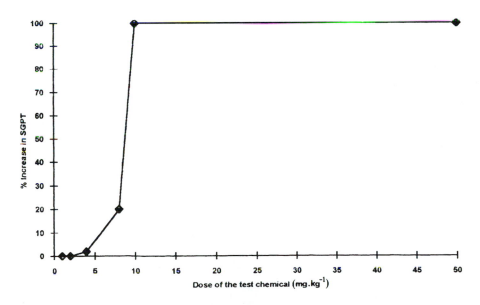

Figure 5.2 Typical results from an acute toxicity test using increased SGPT activity as the endpoint. The NOEL is $2\,mg\,kg^{-1}$ in this experiment.

Plasma

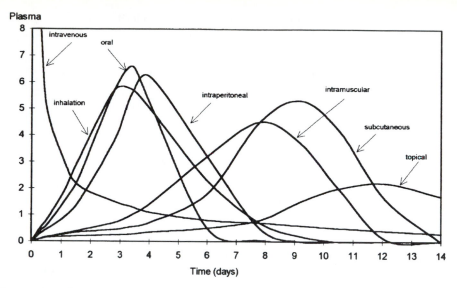

Figure 5.3 Schematic representation of plasma levels of test chemicals following different routes of administration. Intravenous: into a vein; subcutaneous: beneath the surface layer of skin (the cuticle); intramuscular: into a muscle; intraperitoneal: into the peritoneal cavity (this is the abdominal void around the intestines); oral: by mouth, this often means by stomach tube (gavage); inhalation: exposure to an atomised or gaseous form of the test chemical; topical: application to the skin.

to a NOEL or LD_{50} value it is very important to state the route of administration (see Table 5.1) and it is important to select the route of administration in the acute toxicity test which is closest to the route of administration or exposure in humans (e.g. for a pesticide the inhalation route would be appropriate). Similarly the test species has an enormous effect upon the LD_{50} value (see Table 5.2).

The acute toxicity test is carried out by dosing groups of animals with the test chemical, each group at a different dose (including a control group which receives either a placebo, the vehicle (i.e. the solvent for the test chemical) or nothing at all), then observing for deaths in an LD_{50} test or monitoring markers for the NOEL over 14 days. Generally measurements will be taken on a daily basis.

Table 5.1 Some LD_{50} values for cypermethrin, a pyrethroid insecticide, showing the effect of route of administration

LD_{50} (mg Kg^{-1})	Route of administration
250	Oral (in corn oil)
>1600	Dermal
198–315	Intraperitoneal (in dimethylsulphoxide)

Data from *Environment Health Criteria 82*, Cypermethrin, WHO, 1989.

Table 5.2 LD_{50} values for cypermethrin in different test species showing the effect of the test animal upon the apparent toxicity

Test species	LD_{50} $(mg\ kg^{-1})$
Rat	251
Mouse	82
Syrian hamster	400
Chinese hamster	203
Guinea pig	500
Calf	142–284
Piglet	>600
Lamb	283–586

The most difficult question to answer is whether the acute toxicity result for the test species can be extrapolated to humans. In general the order of magnitude of toxicity is extrapolatable, but the exact value is not. If the acute test has been carried out in several unrelated species and all show similar NOELs then it is reasonable to assume that the NOEL can be extrapolated to humans.

5.2.2 *Subacute Toxicity Testing*

Here the effects of multiple doses of the test chemical over a prolonged period (90 days) are determined. The philosophy of the test is similar to that of the acute test; however, here the chemical is dosed every day (usually *ad libitum* in drinking water or feed). Marker enzymes or other parameters (e.g. heart rate) are measured as an endpoint. This is plotted over the duration of the study to determine whether the effects of the test chemical are additive. At the end of the study a full post-mortem examination (PM) is carried out to determine the target organ toxicity (e.g. liver toxicity is often shown by retention of bilirubin and biliverdin in the hepatocytes which results in a mottled yellow appearance of the liver; this is termed icterus). The PM includes histopathological studies to investigate any tissue or cellular microscopical changes that might be associated with administration of the test chemical (e.g. death of liver cells around the microvessels in the liver (centrilobular necrosis) results from exposure to high doses of paracetamol).

As a result of the subacute toxicity study the toxicologist has a great deal more information about the test chemical than was found from the acute test. He is beginning to get a feel for the chemical's effects in animals and is now more able to discuss the potential effects of the chemical upon people. In order to help with the age-old problem of extrapolation from animals to humans, the subacute studies usually use at least two unrelated species (e.g. rat and dog). If the effects are similar between the rat and the dog, both in respect of magnitude of toxicity and target organs, it is reasonable to assume that humans might respond similarly to a dose of the chemical. If not, a great deal of expertise is necessary in discussing the potential effects upon people, and no doubt much argument will ensue.

5.2.3 *Chronic Toxicity Testing*

Chronic means lasting a long time (it is derived from the Greek *chronos*, meaning time). This is the ultimate extension of toxicity testing. In this case the test species is dosed with the test chemical every day for its entire life. It represents the worst exposure case example; if a chemical has no effect in this study then it really is safe. The dosing routes are as previously described (see Section 5.2.1), but usually animals are dosed either orally *ad libitum* or intraperitoneally because these are the easiest dosing routes and a study involving 100 animals dosed every day for 2 years must use simple dosing methods in order to keep staff input at a minimum and therefore costs as low as possible.

We said above that the animals were studied for a lifetime. This is a slight exaggeration. In fact for a rat study a lifetime is deemed to be 2 years. It is important to define a lifetime as just less than the shortest time that we might expect the animal to live under normal conditions so that the study endpoint involves the animals being killed rather than their dying naturally over a protracted time period. Toxicologists like definite endpoints.

The rat is the most common species used in the chronic toxicity study partly because it has a lifespan which is not too long. If dogs were used we might have to wait 10 years for the experimental endpoint. During this time the test chemical's patent would have expired and the company would have no interest in obtaining regulatory approval for its use.

The chronic toxicity study involves much the same protocol as the subacute test. Biochemical and physiological parameters (e.g. body weight) are recorded regularly (e.g. daily) and any dose-related effects are noted. Body weight is a particularly important parameter because it gives an indication of the animal's general health status. An ill rat does not eat and therefore loses weight (see Figure 5.4).

An important purpose of the chronic toxicity study is to investigate the test chemical's carcinogenic potential. The PM carried out at the end of the experiment involves detailed histopathological examination to detect test chemical-associated abnormalities in tissues or cells (e.g. uncharacteristic and rapid growth of a particular cell type, hyperplasia, which might be a precursor of tumour development).

Chronic toxicity tests are exceptionally expensive (of the order of £250 000 per test) and therefore the potential for a chemical to cause cancer (i.e. be carcinogenic) must first be assessed by quicker and therefore cheaper screening methods. There is a series of tests which investigates the potential (i.e. mutagenicity) of a chemical to induce cancer.

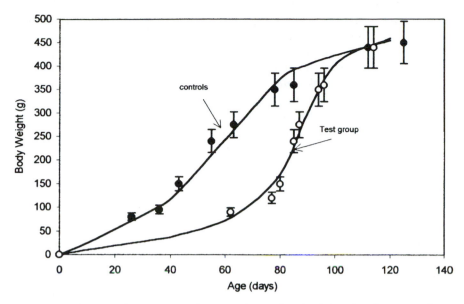

Figure 5.4 Effect of administration of a toxic chemical on body weight of rats in a chronic toxicity study. It is clear by comparison with the controls that the test chemical markedly affects the rat's body weight gain. The test animal, however, eventually reaches the same body weight as the control.

5.2.4 *Mutagenicity Assays*

Results from these tests only indicate whether the test chemical causes aberrations in DNA or RNA which are passed on to daughter cells during mitosis (i.e. mutations). Whether a mutation is expressed as uncontrolled cell growth (a tumour) which might spread (cancer) can only be determined by a chronic toxicity study. As a general rule, all genotoxic carcinogens are mutagens, but not all mutagens are carcinogens. It is important here to distinguish between genotoxic and non-genotoxic carcinogens. It has recently become apparent that some chemicals which cause cancer do so by mechanisms not involving nucleic acids; these are the non-genotoxic carcinogens. Such chemicals would not appear as positives in the mutagenicity assays; the only way that their potential dangers would show up is in chronic toxicity tests. On the other hand, the conventional carcinogens all act by interacting with nucleic acids and so would be expected to show up in the mutagenicity assays. Toxicologists and regulators are getting rather worried about non-genotoxic carcinogens and therefore the requirement for full chronic toxicity testing is becoming greater as more non-genotoxic carcinogens are discovered. For example, sulphadimidine, a sulphonamide antibiotic, is thought to be a non-genotoxic carcinogen which causes thyroid cancer by stimulating thyroid cell growth by directly affecting hormone levels.

The mutagenicity tests investigate the effects of the test chemical upon the DNA (or perhaps RNA) of a specific cell. It is important that there is an in-built mechanism to study the effects of metabolites of the test chemical because many 'carcinogens' (strictly speaking they are pro-carcinogens) need metabolising by the host cells before they are able to interact with nucleic acids and initiate carcinogenesis. For example, benzo[a]pyrene, a very potent pro-carcinogen found in burnt organic materials (e.g. cigarette smoke), is metabolised in the liver (by the mixed function oxidase, cytochrome P_{450}) to form the carcinogen, benzo[a]pyrene-1,2-epoxide.

5.2.4.1   The Ames Test

Bruce Ames is an American scientist who has been working on carcinogenicity for many years. He devised a test that utilises a mutant of the bacterium *Salmonella typhimurium* which is unable to synthesise its own histidine (*S. typhimurium*$^{HIS-}$). It therefore requires histidine in its culture medium. If *S. typhimurium*$^{HIS-}$ is grown on an agar plate containing a very small amount of histidine it will grow for a very short time until it has used up the histidine in its culture medium. If a mutagen is applied to the sparse culture it results in mutations of the bacterial DNA. Some of the mutations will result in a resumption in the ability of the bacterium to synthesise histidine (reversions). The bacterium reverts to the wild type (*S. typhimurium*$^{HIS+}$) and therefore forms colonies on the agar plate. These are termed reversion colonies and are counted and compared with controls (*S. typhimurium*$^{HIS-}$ is also able to revert spontaneously because the HIS$^-$ mutation is in a mutational hotspot on the genome). A significant increase in the number of reversion colonies resulting after exposure to the test chemical constitutes a positive Ames test and means that the chemical is mutagenic.

This test does not account for pro-mutagens (i.e. chemicals which require metabolism to generate the mutagen). For this reason there is a variant of the Ames test which incorporates the supernatant from a centrifuged liver homogenate in the culture medium. The liver homogenate (S9 mix; S9 signifies the centrifugal force necessary to generate the supernatant, centrifugal force is measured in Svedberg units) contains drug metabolism enzyme systems (e.g. cytochrome P_{450}) and so will generate metabolites of the test chemical during the testing period. This test is called the Ames S9 mix test.

Chemicals (e.g. benzo[a]pyrene) which are negative in the Ames test but are positive in the Ames S9 mix test are pro-mutagens. Chemicals (e.g. 1-naphthylamine) which are directly positive in the Ames test are mutagens.

The Ames test relies upon very specific effects upon bacterial DNA and their relevance to mutagenicity and carcinogenicity in mammals is a debate which occupies many a toxicology dinner party. Even Bruce Ames has questioned the importance of a positive Ames test. It is for these reasons that tests based upon mammalian systems have been devised.

5.2.4.2 *The Micronucleus Test*

This test is an *in vivo* test in which the test chemical is administered to the animal (usually a hamster or a mouse), time is allowed for the chemical to reach the target cells and take effect, then the animal is killed and the bone marrow (usually from the femur) is removed and smeared onto a microscope slide, stained and observed under the microscope to check for nuclear aberrations. There are specific nuclear aberrations which are looked for in this test, namely micronuclei in the red cell precursor cells. These small nuclei (or perhaps nuclear fragments) signify DNA damage and point to the chemical being a mutagen.

This test is good because it allows the test chemical to be metabolised by the test animal and so pro-mutagens and mutagens are positive (although it is not possible to distinguish between them). Its major limiting factor is the need for the chemical or its metabolites to enter the bone marrow. In order to determine whether this is possible it is usual to carry out a radiolabelled study in which the animal receives a dose of a radiolabelled form of the test chemical and bone marrow is sampled to see if radioactivity is present. If it is the test is appropriate for that particular chemical. If not it is inappropriate to use the test.

5.2.4.3 *Lymphocyte Chromosomal Aberration Assays*

Strictly speaking these tests investigate clastogenicity (damage to chromosomes) rather than mutagenicity. They are ideal for studies related to humans because human lymphocytes are easy to culture and can be used as the basis of this *ex vivo* test.

Lymphocytes are isolated from blood, cultured and the cultures exposed to varying concentrations of the test chemicals. The cells are exposed to chemicals which terminate the cell cycle at a specific position (metaphase using, for example, colchicine). They are smeared onto a microscope slide, stained and observed under the microscope to see if there are any chromosome abnormalities. The test chemical-exposed cells are compared with controls and the effect of the test chemical is thus determined.

These methods are examples of the many assays designed to detect a test chemical's effect upon the nuclear material. A positive test would necessitate significant further study and possibly would lead to a chronic toxicity test to see if the mutagen is a carcinogen. A single negative test would require a second different test to be carried out; if both proved negative it would probably not be necessary to carry out a chronic toxicity test. This is important because it reduces the number of animals used in toxicity testing and reduces the cost of development of new compounds.

5.2.5 *Reproductive Toxicity Tests*

So far we have considered the direct effects of chemicals upon an individual animal with a view to extrapolating this effect to humans. Reproductive toxicity tests take toxicity testing a stage further and investigate the effects upon the offspring of exposed animals.

Reproductive toxicity tests answer one of the following questions:

1.   What is the effect of dosing the pregnant mother?
2.   Does exposure of the female to a particular chemical before she becomes pregnant cause changes to her reproductive apparatus which result in malformed offspring?
3.   Does exposure of the male result in sperm damage which results in malformed offspring?
4.   Does crossing an offspring from a treated mother with another animal result in malformed second generation offspring?

These are all issues that reproductive toxicity studies must address; however, most new compounds are not studied in this intense detail unless there is reason to believe that they have reproductive toxicity potential.

It is beyond the scope of this book to cover reproductive toxicity testing in detail. For this reason only teratogenicity studies are covered. The word teratogenicity is derived from the Greek, *teratos*, meaning monster, and genesis – to generate a monster. This is a very appropriate term for the study of birth defects. It refers to chemicals which damage the developing embryo directly (e.g. by damaging DNA and so preventing cell division) rather than having an effect upon the dam which affects her physiology (e.g. altering glucose uptake) which might indirectly affect the embryo.

Teratogenicity studies involve the administration of the test chemical to the pregnant female (dam) during the course of her pregnancy. The young (pups) are delivered by caesarean section 1 day pre-term. They are observed for defects such as the presence of extra ribs, cleft palate or biochemical defects. Bone defects are visualised by clearing the foetus (or embryo if termination is at an earlier stage in pregnancy) with potassium hydroxide and staining the bones with alizarin.

Rodents are commonly used for teratogenicity studies (the rat is the commonest) and they usually resorb damaged embryos rather than miscarrying them. Why miscarry when you can resorb valuable nutrients? It is important to look for signs of resorption because embryotoxicity might not result in the delivery of a damaged foetus. For each resorbed embryo there will be a resorption site on the wall of the uterus. This is a small dark spot which the expert can easily see. Thalidomide, the classic teratogen, often does not result in the delivery of deformed offspring in rodent studies, but rather results in reduced litter sizes and an increased number of resorption sites compared with controls. Clearly it is very important indeed to monitor both parameters in the teratogenicty studies for fear of missing another teratogen.

5.3   Interpreting Toxicity Data

It takes at least 2 years to complete a full toxicity study and at its end there is a large amount of data. Sometimes the toxicity data for a new molecule can fill six or seven 200 page volumes. The toxicologist must interpret these data in order to determine the risk versus benefit of using the compound. Consider, for example, a drug for the treatment of the common cold. If such a drug demonstrated tumours in the chronic toxicity test at a dose level near to the drug's intended dosing level, no toxicologist would entertain its approval for use in human medicine. The risk far outweighs the benefit. On the other hand, if the drug were for the treatment of a severely life threatening disease (such as cancer), its risk, albeit high, might be acceptable in the context of the benefits of possible cure or extension of life.

In the UK, the Committee on the Safety of Medicines (CSM; an independent committee of experts which advises the Secretary of State for Health) decides whether a licence should be granted to permit the use of a proposed medicine. The CSM weighs up the toxicity data, therapeutic benefits and social implications of the new medicine in coming to its decision. If the drug were to be used as a veterinary medicine a similar committee, the Veterinary Products Committee (VPC), reviews the data and advises the Ministry of Agriculture, Fisheries and Food and Health whether a licence should be granted. There are environmental implications associated with both veterinary and human medicines. Having taken, or been administered, a medicine, and its metabolites are excreted and either flushed down the toilet (in the case of human medicines) or deposited onto fields or into slurry lagoons (veterinary medicines). The effect of these upon ecosytems is now an important consideration in the licensing of both human and veterinary medicines.

Environmental contamination following the use of veterinary medicines (and to a very much lesser extent human medicines) might result in residues in food. More importantly, the use of veterinary medicines is likely to result in residues in meat. It is important to consider the potential effects of these sources of dietary contaminants upon human health.

5.4   Human Exposure to Environmental Chemicals

People can be exposed to environmental pollutants via several routes:

1.   In food (e.g. residues of pesticides used in agriculture)
2.   In the workplace (e.g. chemicals used in industrial processes)
3.   Direct from the environment (e.g. breathing polluted air in cities).

5.4.1   *Food as a Source of Exposure to Environmental Pollutants*

5.4.1.1   *Pesticides*

Pesticides are important in modern food production; without them it is unlikely that we would be able to produce sufficient food to fulfil the enormous world market.

As it is, a good proportion of the world's population is starving (politics are an important reason for this, but we cannot go into that here) even with a battery of pesticides designed to kill most of the pests that most crops would ever encounter. When pesticides are used they might leave residues in the crops upon which they have been used and these residues might find their way into the foods produced from the crops (e.g. residues of pesticides used in wheat production might appear in bread). It is the concentration (i.e. risk) of the residues in the foodstuff that is most important rather than just the fact that they are there (i.e. hazard). Maximum residue levels (MRLs) are set to safeguard the consumer against residues levels which might be of toxicological significance. Having said that, it is important to remember that pesticide MRLs are set on the basis of good agricultural practice (GAP); the MRL is the maximum concentration of a pesticide that will occur in a crop if it has been used properly by the farmer. Levels above the MRL are not permitted. There is no direct toxicological consideration in this calculation; however, if there is toxicological concern about a value derived from the GAP method it is very likely that the MRL will be modified. We can only hope that in time the pesticide MRL calculation will include consideration of the toxicity of the pesticide. The reason for this last statement is illustrated well when veterinary medicine MRLs are considered (see below) because here toxicity is an important consideration. It is therefore possible to generate the ludicrous situation that a single pesticide used in both crop protection and veterinary medicine (e.g. Diazinon, an OP, used in sheep dips and crop protection) could have two MRLs, one following its use as a medicine, the other following its use as a crop insecticide.

The UK Government's Working Party on Pesticide Residues (WPPR) monitors the human food chain for pesticides; in 1995 it found that only 1% of food contained residues above the MRL. There were no cases which would have resulted in the ADI being exceeded. It is important not to overshadow the issue with perceived risk here; the UK press would make a meal of a tiny proportion of the human food chain containing pesticides whereas a family in Ethiopia would simply be glad of something to eat. We must keep this issue firmly in perspective.

There are several examples of residues of pesticides in food intended for human consumption. One of the best of these occurred in the UK in 1990 and involved the OP isofenphos. Isofenphos is used as a seed dressing (i.e. crop seeds, such as wheat, are coated with the OP to prevent their being eaten by insects when planted) in several countries within the EU (it is particularly toxic and so is not licensed in the UK). Early in the summer of 1990 a large number of pigs on UK farms began to show signs of ataxia (unsteady gait). Intensive study showed that their meat contained residues of isofenphos which was responsible for the ataxia. The meat was not allowed to go into the meat market because of the possible toxic effects of isofenphos upon the consumer. Detailed study demonstrated that the isofenphos had originated in a French warehouse where a component of the pigs' feed had been stored near to isofenphos-dressed grain. Cross-contamination had occurred and resulted in the pigs' diet being contaminated (see Figure 5.5). This is an excellent example of a food chain residues problem which was, fortunately, identified early and rectified before the human consumer was put at risk.

CONTAMINATION IN WAREHOUSE (18.3 mg kg^{-1} in dust from floor)

⇓

WHEAT SCREENINGS (156 mg kg^{-1})
(Waste product from grain industry used in animal feeds)

⇓

PIG FEED (1.8 mg kg^{-1})

⇓

PIG ADIPOSE TISSUE (0.01 mg kg^{-1})

Figure 5.5 Route by which isofenphos, an OP, contaminated pig tissues in the UK in 1990 showing the isofenphos concentrations at each point. Data from I.C. Shaw *et al.*, 1995, *Vet Record*, **136**, 95.

5.4.1.2 *Veterinary Medicines*

The situation discussed earlier in relation to pesticides is mirrored by veterinary medicines. Medicines used either to treat farm animals' diseases, as growth promoting agents or in the prevention of disease (prophylaxis) might find their way into the meat derived from the treated animals.

The MRL situation for veterinary residues is very different (as discussed earlier) to that for pesticides; MRLs for veterinary residues are derived by a calculation which includes a consideration of their toxicity to the consumer.

We have already discussed the problem with clenbuterol residues in beef (see Section 2.2.7) which illustrates one of the routes by which veterinary medicines might form residues in meat intended for human consumption. There are many more examples. One of the most topical is the ectoparasiticide dichlorvos, an OP used in the treatment of lice in farmed fish (trout and salmon). The market for such fish is increasing and is an important facet of the economies of countries such as Norway, Canada and Scotland. Residues of dichlorvos might occur in fish intended for human consumption, but perhaps more importantly dichlorvos might have a significant and irreversible effect upon the aquatic environments (e.g. lochs and fjords) in which fish are farmed. This is of great concern at the moment.

Legislation can have an enormous impact upon veterinary residues. For example, the use of hormone growth promoters was banned in the then EEC in 1981. This had an immediate effect upon residues of hormones in meat (see Figure 5.6); it is now very rare indeed to find such residues in meat in the EU. The situation in the USA is quite different because growth-promoting hormones are still permitted.

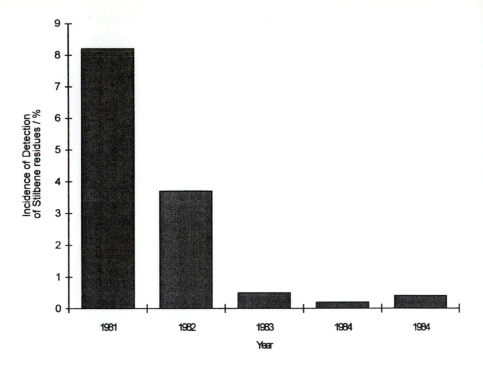

Figure 5.6 Decline in residues of diethylstilboestrol (a steroid growth promoter) in meat in the UK following its ban in 1981. Data from Ministry of Agriculture, Fisheries and Food, UK.

5.4.2   *Exposure to Toxic Chemicals in the Workplace*

It is important to consider humans as part of the environment and consequently the workplace can be considered as an ecosystem. Entire books have been written on the diseases caused by occupational exposure to chemicals (the first was by Hunter in 1950) and therefore a short section in a book such as this is inadequate. There are, however, several classic examples which illustrate the principles well.

5.4.2.1   *Scrotal Cancer and Chimney Sweeps*

Perhaps the first example of an occupational disease due to exposure of workers to toxic chemicals was scrotal cancer. In the nineteenth century boys were often employed to climb up chimneys and clean them *en route*. During their work they were exposed to many carcinogenic polycyclic aromatic hydrocarbons (e.g. benzo[a]pyrene) and this, combined with their questionable personal hygiene, resulted in a greater incidence of scrotal cancer in chimney sweeps than in other members of the population of the time.

5.4.2.2 *Naphthylamines*

1-Naphthylamine (see Figure 5.7) is used in the manufacture of dyes. In the 1960s it was realised that workers in the dye industry had a higher incidence of bladder cancer than the general population. A brilliant epidemiological study (one of the first of its kind) linked the bladder cancers to exposure to 1-naphthylamine. Later work showed that 1-naphthol was at best a very mild carcinogen and that the reason for its apparent carcinogenicity in the workplace was due to impurities of 2-naphthol which is one of the most carcinogenic chemicals known.

Changed work practice resulted in reduced exposure to the carcinogenic naphthylamines and resulted in the incidence of bladder cancer in dye workers returning to that expected for the general population.

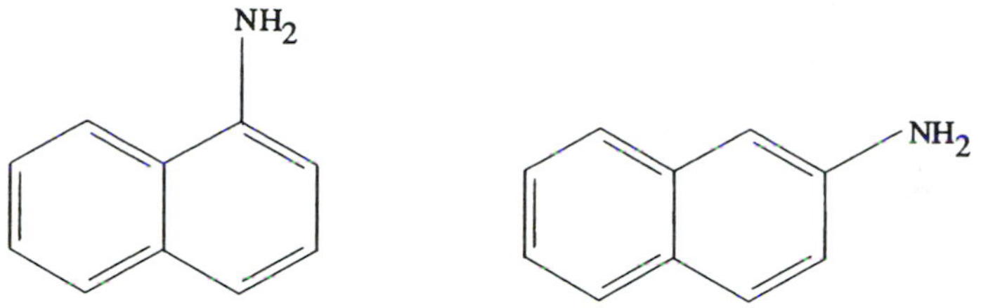

1-Naphthylamine   2-Naphthylamine

Figure 5.7 Molecular structures of 1- and 2-naphthylamine. The former is used in the manufacture of dyes. The latter is a profoundly carcinogenic impurity thought to be responsible for the high incidence of bladder cancer seen in workers in the dye industry in the 1960s.

5.4.2.3 *Nickel Fumes*

Nickel is an unusual element because it is carcinogenic (as is chromium, both likely to be toxic because of non-genotoxic mechanisms). Nickel smelters or men who weld using nickel are exposed to nickel fumes and have an increased incidence of nasal cancer. Protection of such workers against the carcinogenic vapour is now mandatory.

5.4.3 *Direct Exposure to Environmental Chemicals*

There are many ways in which one might be directly exposed to environmental pollutants. Perhaps the two most important are via drinking water and by walking in countryside where pesticides have been used.

5.4.3.1 *Drinking Water*

Drinking water is protected by legislation which is strictly policed (see Chapter 8); however, there are notable examples of contamination of ground water that have resulted in contaminated drinking water. One such case occurred in Camelford, Cornwall, in the UK in 1985. An aluminium salt used to clarify water as part of the normal water purification process was emptied by mistake into the wrong reservoir at the Camelford water treatment plant. The result was a high concentration of Al^{3+} in the drinking water of the residents of the area. Arguments continue as to whether the people have suffered ill effects, but clearly there are worries as aluminium is thought to be associated with the development of presenile dementia (Alzheimer's disease).

In another case the use of atrazine (a herbicide) has resulted in its appearance in drinking water (see Section 9.2). This is particularly worrying because it is a suspect carcinogen. Atrazine's use in the EU is now severely restricted in order to prevent the continuation of the problem.

5.4.3.2 *Walking in the Countryside*

If you walk through a field that has recently been sprayed with a pesticide you will be contaminated. In the UK there is no requirement to place notices at the entrance to such fields, although some public-spirited farmers who have public footpaths running through their fields do warn walkers of the hazard. It is, of course, unlikely that exposure in this way will result in signs of toxicity; however, dogs and cats can die as a result of exposure to pesticides in this way.

Further Reading

Klaassen, C.D. (Ed.), 1996, *Casarett and Doull's Toxicology — the Basic Science of Poisons*, 5th Edn, New York: McGraw-Hill.

Timbrell, J.A., 1991, *Principles of Biochemical Toxicology*, 2nd Edn, London: Taylor & Francis.

6

Fate and Behaviour of Chemicals in the Environment

This chapter discusses the behaviour and disposition of pollutants in the environment. The physicochemical properties of pollutants are discussed in the context of how they interact with (for example) soil:

- Log octanol/water partition coefficient
- Hydrolysis
- Photolysis
- Adsorption/desorption
- Mobility
- Volatility

Interpretation of data is discussed in assessing persistence, bioaccumulation, bioconcentration and biomagnification.

Breakdown of pollutants in the environment is discussed and the major xenometabolic pathways introduced:

- Cytochrome P_{450}
- Glucuronidation
- Sulphation
- Amino acid conjugation
- Glutathione conjugation
- Bacterial metabolism

6.1   Distribution of Chemicals in the Environment

Within the field of environmental toxicology we are often placed in a situation of having to predict the probability of exposure (i.e. risk) to environmental chemicals from a very limited scientific database. After the release of a chemical into the environment we often know little of its fate and behaviour in the different compartments of the environment (e.g. air, soil/sediment, water, biota).

We know even less about the transformation or detoxification processes involved in the vast majority of chemicals in the environment, although an extensive body of literature exists on the fate and behaviour of extensively studied chemicals such as DDT, tri-organotins and some organochlorines.

Chemicals which are deliberately applied to, or escape into, the environment move from a source or input point to eventually end up in one or more compartments of the environment depending on their toxicity, physicochemical properties, persistence and mobility. In the process most chemicals are transformed by biological, physical and chemical processes into a number of different breakdown products (if a biological process is involved these are metabolites). In most cases we can assume the process over time results in the complete breakdown of a substance resulting in detoxification, thus rendering substances harmless to living organisms. This is the ideal situation, but it is possible that the detoxification process might result in the formation of products that are more toxic (e.g. *g*-hexachlorocyclohexane is biotransformed to a free radical which is very toxic). We have already mentioned in previous chapters the toxic effects of tri-organotins (TOTS) on the dog whelk and other aquatic life. TOTS have been used in wood preservatives and antifouling products since the 1960s. Had we known of their toxic effects and fate and behaviour earlier they would probably never have been approved in the first place. TOTs are compounds that are characterised by three tin–carbon covalent bonds (see Figure 6.1). They are lipophilic, have low aqueous solubility and adsorb strongly to soil and sediment. They therefore have a tendency to persist in the environment and build up in plant and animal tissues. TOTs can persist in aerobic soils and sediments for up to 815 days whereas their persistence in water is often less than 5 days. Degradation and transformation occur through a mechanism of sequential debutylation (i.e. loss of the butyl side chains). The most significant degradation processes are microbial degradation and photodegradation. Although the process of debutylation inexorably leads to the formation of inorganic (very much less toxic) tin in the environment, the rate of transformation varies greatly between different environments.

The main abiotic factors that seem to enhance the degradation process are elevated temperature, increased sunlight and aerobic conditions. The fate and behaviour of chemicals in the environment are not simple. Many environmental

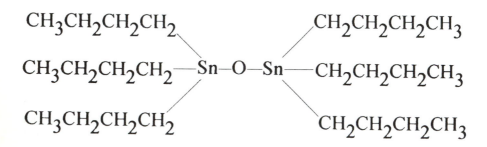

Figure 6.1 Molecular structure of tributyltin oxide (TBTO)

factors will affect these processes and we must be able to understand these in order to carry out an effective risk assessment. The assessment of the extent of chemical pollution and the impact of the chemicals' fate and behaviour is a sophisticated science. Different chemicals have different physicochemical properties and are used in different ways. There is a number of ground rules which must be observed when attempting to predict the fate and behaviour of any one individual chemical in the environment. By far the most important consideration is that of the physicochemical properties of the chemical.

6.2   Physicochemical Properties

There is a number of physical and chemical properties which will affect the fate of a chemical in the environment. As a general rule it is worth remembering that in air, soil, sediment and water the behaviour of ionic pesticides and those with ionisable chemical groups differs from that of neutral and non-polar molecules. The polarity of a molecule is determined by the position of electrons in its molecular structure. A polar molecule has a charge separation which means that one end is positive and the other negative (i.e. electron rich). This charge separation does not necessarily result in an absolute charge, but may only involve a relative charge difference. The latter is denoted as δ-charge (eg. δ-ve). Polar molecules (which are hydrophilic, from the Greek *hydros* meaning water and *phillos* meaning to like) are more water soluble than apolar molecules (which are hydrophobic, *phobos* means to hate; they are sometimes termed lipophilic meaning they like fats) (see Figure 6.2). The movement of electrons in a molecular structure is influenced by the molecule's constituent atoms. Some elements have electron-withdrawing properties (they are electronegative) and therefore aid polarity. The electronegative atoms are fluorine, chlorine, oxygen, nitrogen and sulphur in descending order of electronegativity. The more non-polar or lipophilic a compound, the more likely it is that it will prefer an organic-rich site in which it may be strongly adsorbed. Listed below are the main physicochemical properties that affect the fate and behaviour of a chemical in the environment:

- Molecular weight
- Melting point/boiling point
- Vapour pressure
- Solubility in water
- Partition coefficient between water/sediment, water/soil and water/natural lipids.

All of the above information and subsequent data are readily available by conducting standard tests to OECD (see Chapter 8) guidelines. The importance and use of the above properties can be explained simply as follows. High water solubility limits the vapour loss of a pesticide from the soil to the atmosphere, despite the volatility of the chemical substance. Water solubility will also indicate the likely mobility of a chemical through a soil profile (i.e. the layers of soil), and

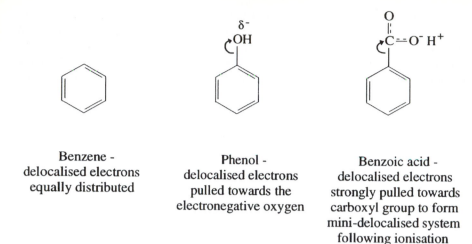

| Benzene -
delocalised electrons
equally distributed | Phenol -
delocalised electrons
pulled towards the
electronegative oxygen | Benzoic acid -
delocalised electrons
strongly pulled towards
carboxyl group to form
mini-delocalised system
following ionisation |

HYDROPHOBIC ⎯⎯⎯⎯⎯⎯⎯⎯⟶ **HYDROPHILIC**

Figure 6.2 Benzene, phenol and benzoic acid showing differences in the positions of their electron charge clouds giving rise to differences in polarity. Benzene is very apolar and therefore almost water insoluble, phenol has δ-charge and so is more water soluble whereas benzoic acid is ionised and is very water soluble.

can be used to predict the likelihood of a chemical reaching groundwater. Data on the volatility/boiling point will indicate if a chemical is likely to stay on the soil surface and can also be used in the prediction of mobility. The log octanol water partition coefficient can be used to calculate how strongly a chemical is bound to soil/sediment particles but, more importantly, it is used to predict if a chemical is likely to bioaccumulate in plant and animal tissue. In fact the partition coefficient (log P_{ow}) of a substance can be considered the single most important physico-chemical parameter in predicting the fate and behaviour of a chemical in the environment. Partition coefficients will include those for n-octanol/water, soil/water, sediment/water and air/water.

6.2.1 The n-Octanol/Water Partition Coefficient (log P_{ow}; Hansch Coefficient)

The lipophilicity of a molecule is measured by determining its relative solubility in water and a long-chain hydrocarbon. If the molecule prefers water it is unlikely to concentrate in cell lipids, if it prefers the organic solvent it is likely to seek out the lipids. The procedure was refined by a German scientist, Hansch, working in the USA in 1963. He used octan-1-ol ($C_8H_{17}OH$) to represent lipids. The method is simple. Equal volumes of octan-1-ol and water are mixed in a separating funnel, the test chemical is added and the whole shaken vigorously. The mixture is allowed to stand to allow the aqueous and organic layers to separate. The concentrations of the test chemical in both the octan-1-ol and water are then determined (e.g. by gas-liquid chromatography, see Chapter 4). The Hansch Coefficient is then calculated:

$$\text{Hansch Coefficient} = \log_{10} \frac{[X]_{octanol}}{[X]_{water}}$$

where X represents the test chemical. The Hansch Coefficient is now usually termed the log octanol water partition coefficient (log P_{ow}).

Log P_{ow} is a very important parameter because we know that if it is greater than 3 (i.e. the test chemical is 1000 times more soluble in octan-1-ol than in water) then the chemical is very likely to concentrate up the food chain. A value of 3 is therefore the cut-off point. Chemicals with log P_{ow} > 3 are unlikely to be licensed unless very good evidence for their lack of long-term environmental toxicity is also presented (see Table 6.1 which shows log P_{ow} values for several well-known pesticides).

The 'solvent shaking method' is the traditional way of determining log P_{ow}; however, it is now often determined using high performance liquid chromatography (HPLC; see Chapter 4). The retention time on a reverse-phase HPLC column (the column has bonded C_8 (octane) carbon chains which is, in effect, immobilised octane: the mobile phase is aqueous based) is a function of the differential solubility of the test chemical in octane and water and is used to calculate log P_{ow}.

As a predictive tool there are serious implications if log P_{ow} is greater than 3. In this case the chemical is likely to be tightly bound to organic matter especially fat. It is therefore highly likely that a chemical will bioconcentrate, bioaccumulate or biomagnify within the environment. These three processes are explained below.

Table 6.1   Log P_{ow} values for a selection of well-known pesticides

Pesticide	Log P_{ow}
DDT	6.2
Cypermethrin	6.3
Tributylin oxide (TBTO)	3.2–3.8[a]
Atrazine	2.6

[a] This is the range of values for TBTO compounds generally.
It is clear that pesticides which are known to cause chronic environmental harm have values very much greater than 3. Even though cypermethrin has a value greater than 3 it is rapidly hydrolysed on contact with water and so does not persist.

6.2.1.1 *Bioconcentration*

This is a process in which compounds or substances enter organisms directly (e.g. via gills) and concentrate in tissue.

6.2.1.2 *Bioaccumulation*

This includes bioconcentration and also the uptake of residues through the food chain.

6.2.1.3 *Biomagnification*

This refers to the total process in which tissue concentrations of a lipophilic chemical increase through two or more trophic levels.

There is a very clear relationship between log P_{ow} and bioaccumulation which appears linear from log 3 up to around log 6. It is interesting that we may have been able to predict better the fate of DDT in our environment had we observed its log P_{ow}. In the case of DDT its persistence and bioaccumulative properties had a disasterous effect on birds of prey. Its log P_{ow} value is approximately 6. Log P_{ow} values below 3 rarely result in bioaccumulation, but anomalies do exist depending on (for example) species physiology. For instance, certain marine molluscs lack the enzymes to break down some chemicals, hence they unexpectedly build up in body tissue. Used in conjunction with other simple tests, log P_{ow} can be a useful tool in predicting the fate of chemicals in the environment. It can also save extensive animal testing and the unnecessary use of laboratory animals and wildlife.

6.3 Prediction of Fate and Behaviour

Linked to the simple physicochemical properties of a substance are the transformation and behaviour studies that are needed to more accurately predict the fate and behaviour of a chemical in the environment. With a few additional simple tests in conjunction with the physicochemical data it is often possible to go a long way to predicting the eventual fate and behaviour of chemicals in the environment. These additional tests are as follows.

6.3.1 *Hydrolysis as a Function of pH*

The chemical and physical stability of the test chemical is very important in assessing its potential effects upon the environment. If the molecule is denatured immediately upon contact with water it is likely to have a very low environmental impact. It must be remembered, however, that the degradation products might be equally as toxic, or even more toxic, than the parent molecule.

Simple hydrolysis experiments are carried out to assess the stability of a test chemical in an aqueous ecosystem. The chemical is added to water buffered to different pHs so that an assessment of its stability to different environmental conditions can be made (to illustate differences in environmental conditions and the importance of pH, an upland peaty stream might be pH 4.5, whereas a limestone brook might have a pH of 9). After incubation for several hours the solution is analysed to see if the test chemical's concentration has decreased and whether any degradation products have appeared.

There are specific chemical groups which confer low water stability. The most important of these is the ester ($-C(O)-O-CH_2-$) which is readily hydrolysed to an alcohol ($-CH_2-OH$) and a carboxylic acid ($-COOH$). The pyrethroids (an important group of pesticides) have an ester bond (see Figure 2.11) in their molecular structures and readily hydrolyse in aquatic environments. It is for this reason that the environmental impact of pyrethroids is lower than might be predicted from the environmental toxicity test results (they are extremely toxic to fish, but their rapid hydrolysis when used as agricultural insecticides means that they do not reach waterways intact and therefore fish are not severely affected by their use).

6.3.2 *Adsorption/Desorption*

The test chemical is added to different soil types and the amount of free and bound (i.e. adsorbed) chemical is determined. Chemicals with high log P_{ow} values tend to adsorb better.

6.3.3 *Photolysis Studies*

Stability to ultraviolet (UV) light is another important consideration. This is particularly important in assessing the potential environmental toxicity of pesticides because when they are sprayed onto plants they form a thin film on the leaves and stems. This film is exposed to sunlight and if the pesticide is UV labile it will quickly break down. As part of the physicochemical studies, stability to UV light is assessed. The chemical is exposed to light (at different wavelengths) in aqueous solution and the rate of degradation is determined.

6.3.4 *Degradation in Natural Water/Sediment*

The test chemical is added to sediment plus water from a river or stream. The rate of degradation (i.e. half-life, $t_{1/2}$) is measured.

From the combined information it is possible to predict both the mobility and persistence of a chemical in the environment along with details of any transformation products which may be more or less toxic/stable than the parent compound.

6.4 Mobility

Mobile chemicals (those likely to leach from soil into water) tend to have a water solubility greater than 30 mg dm^{-3}. They will also have a soil/water distribution coefficient (K_d) of less than 2. K_d is determined by mixing soil, water and the test chemical together and measuring the concentration of the chemical in the aqueous phase:

$$K_d = \frac{[\text{chemical in aq. phase}]}{[\text{total chemical}] - [\text{chemical in aq. phase}]}$$

Another way of predicting the mobility of a chemical in an ecosystem is to determine its binding to organic material (e.g. humus). This is termed the organic carbon binding constant (K_{oc}). K_{oc} is calculated from K_d, where:

$$K_{oc} = \frac{K_d}{\%\text{ organic carbon in sample}}$$

The organic carbon content of the soil sample is determined by combustion techniques. The difference in mass before and after heating to high temperature corresponds to the organic carbon content.

Mobile chemicals have K_{oc} values of less than 500. In addition, mobile chemicals are often characterised by a Henry's Law Constant (K_H) of less than 10^{-2} atm m^3 mol^{-1} (i.e. the chemical is volatile):

$$K_H = \frac{[\text{test chemical in water}]}{[\text{test chemical in air}]}$$

The ionic forms of mobile chemicals tend to be negatively charged. Negative ions are more mobile in soil and silt systems because they are repelled by negatively charged clay particles; positive ions are attracted and therefore their passage is retarded. The pH is important here too because at low pH (i.e. high [H$^+$]) the clay (or other negatively charged) binding sites are blocked by H$^+$. This means that compounds are more mobile in acid conditions (see Figure 6.3).

6.5 Persistence

Persistence along with mobility can be considered the two most important factors that dictate the effects a chemical can have on the environment. Like mobility, persistent chemicals are characterised by specific properties which can be predicted from physicochemical data and fate and behaviour studies. Persistent chemicals normally have a half-life of greater than 25 weeks in a hydrolysis test as a function of pH. Their persistence is governed by the rate of hydrolysis, hence a

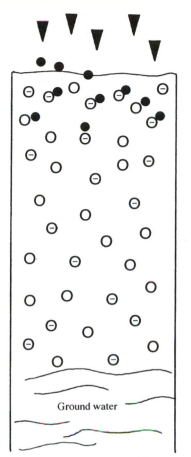

Positively charged molecules (●) will be electrostatically attracted to the negatively charged soil particles (⊖).

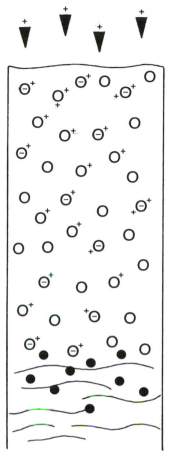

pH change (towards acid +) can either displace positively charged contaminants from their electrostatic interaction with soil particles or alter the molecule's intrinsic charge (dependent upon pKa value). This results in the once bound contaminant leaking through the soil into groundwater.

Figure 6.3 Schematic representation of the flow of molecules through soil (or silt), showing the effects of pH and the intrinsic charge of the molecules.

chemical tends to be less persistent in moist warm soil. Persistent chemicals tend to have a half-life in soil of greater than 2–3 weeks and greater than 1 week in photolysis studies. The rate of degradation and hence persistence, or otherwise, is dependent not only on the intrinsic properties of the chemical but also on the biotic and abiotic properties of the soil, whether a chemical is in soil, sediment or water, whether in sunlight or shade, acid or basic pH and the organic carbon content of the soil. Perhaps surprisingly, mobility and persistence data derived from these simple tests are reliable indicators of the fate and behaviour of chemicals in the environment. The UK Government has devised a classification scheme for regulatory purposes to facilitate the assessment of the potential of a particular pesticide reaching specific environmental compartments (e.g. groundwater; see Table 6.2).

A further tool adopted by regulators is the groundwater ubiquity score (GUS). Based on environmental fate and behaviour properties, it is used to predict the potential of a chemical to enter groundwater:

$$GUS = (\lg \text{ soil } t_{1/2}) . (4 - (\lg K_{oc}))$$

A high GUS value indicates that the chemical is not likely to enter groundwater.

Table 6.2 Classification of pesticides according to their mobility and persistence in soil

Mobility class	K_{oc} $(cm^3 g^{-1})$	Persistence class	$t_{1/2}$ (days)
Non-mobile	> 4000	Non-persistent	< 5
Slightly mobile	4000−500	Slightly persistent	5−21
Moderately mobile	499−75	Moderately persistent	22−60
Mobile	74−15	Very persistent	> 60
Very mobile	<15		

6.6 Adsorption/Desorption Properties

Adsorption/desorption is another important process which affects the fate and behaviour of a chemical in soil, water and sediment. Chemicals which are strongly bound to soil or sediment will not be mobile except by physical transportation (e.g. soil being washed from one place to another by heavy rainfall) and will not be available for uptake by organic organisms (low bioavailability).

The factors that affect whether a chemical is bound or released from soil are particle size, type of soil (e.g. sandy or clay), organic matter, cation exchange capacity, water solubility and the octanol water partition coefficient. Any one or a combination of these will affect the fate and behaviour of a chemical in the environment. For instance, the presence of dissolved organic carbon in the water phase will reduce the adsorption of lipophilic chemicals like DDT onto sediment (because the dissolved organic molecules will occupy a proportion of the binding sites). The volatilisation or loss of a chemical to air sometimes complicates the issue further. For instance, some chemicals are more volatile than others thus to ascertain their fate in soil and water one needs to know the amount of a chemical lost to the atmosphere; this can be determined experimentally.

6.7 Volatility

An important consideration in assessing environmental impact is the chemical's ability to vaporise. If it is very volatile it is unlikely to have a major impact upon the terrestrial environment because it will quickly be vaporised by air movement. If the molecule has a low water solubility and is volatile it is unlikely to have a significant impact upon aquatic environments because it will form a surface film that will evaporate quickly. Volatility is measured as vapour pressure (Vp); this is the saturation pressure above the test chemical in a sealed container at a specific temperature (usually 25°C). High Vp means high volatility (see Table 6.3).

Table 6.3 Vapour pressures (Vp) and boiling points (BP) of several pesticides and other volatile chemicals showing their approximate correlation

Chemical	Vp^a (Pa)^b	BP (°C)
Ethoxyethane (diethyl ether)	434	34
Pentane	422	36
Ethanol	42	78
Benzene	75	80
Octane	10	125
Diazinon	1.4×10^{-4} mmHg	83−84
Parathion	3.8×10^{-5} mmHg^c	157−162

[a] Saturation vapour pressure at 25°C unless otherwise stated.
[b] Pa = Pascal = $N\,m^{-2} = 7.5 \times 10^{-3}$ mmHg.
[c] Vp at 20°C

6.8 Interpretation of Physicochemical Properties Data

Even armed with all our knowledge of physicochemical properties, persistence and mobility data and partition coefficients, our predictions of fate and behaviour are still at best tenuous. We have not taken account (yet) of the biotransformation of the chemical and cannot include in our models the enormous physical variations (e.g. climatic conditions) which occur in real ecosystems. We can, however, with certainty predict three possible outcomes of the fate and behaviour of a chemical in the environment:

- the chemical remains where it originally entered the environment
- the chemical is carried elsewhere into soil, sediment, water or atmosphere
- the chemical is transformed or broken down by chemical, physical or biological processes.

The assessment of the likely extent of any of the above eventualities in the overall fate of the chemical in the environment is termed the mass balance. Many model systems have been devised to calculate mass balance. This is done in order to predict accurately the fate and transport of a chemical or its breakdown products in different compartments in the environment. Using the data derived from techniques we have described previously, it is possible to construct a mass balance equation. Most models start with the assumption that all chemicals and their breakdown products will reach steady state in the environment and will eventually reach a final sink in one compartment of the environment. Models assume an initial source of a pollutant or chemical and predict a final sink. A necessary parameter in a mass balance calculation is the control volume (e.g. the volume of water/sediment in a lake) which from within its boundaries we can calculate the total transport of a chemical, its breakdown and final sink.

Few models are simple and even fewer have been validated. Many grossly overestimate or underestimate the level of pollution in different compartments of the environment and few have any great precision or accuracy. The art of modelling of fate and behaviour is at present exactly that, an art. Many environmentalists will make great claims for particular models and regulators swear allegiance to methods not involving excessive costs. It is difficult without the model makers to make any more than a qualitative subjective estimate of the fate of chemicals in the environment. Modelling a chemical's fate in the environment could be the subject of a completely separate text and has been aired more completely by authors elsewhere. Suffice to say the body of knowledge developed on the fate and behaviour of chemicals is ever expanding and model systems are becoming ever more sophisticated. We are fast becoming able to predict from simple tests and calculations the eventual fate and behaviour of chemicals in the environment. The predictive element in fate and behaviour assessment precludes the necessity for post-approval monitoring and clean-up campaigns after environmental damage. We are getting it right now more often than we are getting it wrong and the fate of chemicals in the environment is left less to the vagaries of the security of a dilution factor and more to the practised knowledge of regulators and scientists with a vested interest in the health of the environment.

6.9   Degradation of Molecules in the Environment

When a molecule enters the environment it is subject to two degradative forces: physical (e.g. UV light) and biological (e.g. microbiological). We now turn our attention to the biodegradation of molecules in the environment.

There are 10^7 species of plants and animals in the biosphere and each is potentially capable of modifying a molecule with which it comes into contact. Very little is known about the xenometabolic pathways of the vast majority of animals and plants, but it is well known that there are very great species differences in the way in which animals (and presumably plants, although very little is known about the xenobiochemistry of plants) deal with xenobiotics. In considering the biological degradation of environmental pollutants it is important to consider the enormous array of species that might be involved in degradation in a single ecosystem and to bear in mind that there are very many metabolic pathways by which a xenobiotic might be metabolised. In this chapter we consider the general classes of metabolic pathways which might be involved in the degradation of environmental contaminants.

6.9.1 *Species Differences in Metabolism*

Species differences in metabolism are particularly important in determining the effects of chemicals upon individual animals and plants rather than their routes of degradation in a particular ecosystem. This is because there is such a large number of individual species in any one ecosystem that the overall metabolism of a contaminant in that ecosystem may be regarded as an average of all the degradative efforts of the individual organisms. It is therefore very likely indeed that a species will not have a profound effect upon the degradation of a contaminant of an ecosystem of which it is a member. On the other hand, if an individual species that is unable to degrade (detoxify) a particular environmental contaminant absorbs that contaminant, even though it is unable to metabolise it, it is very unlikely to affect the overall fate of that molecule in the ecosystem. The individual animal (or plant) that absorbs the chemical and is unable to detoxify it will, however, suffer the toxic effects of that chemical. Species differences in metabolism are therefore important in determining the impact of a chemical upon the environment, but not the impact of the environment upon the chemical.

The importance of species differences in metabolism is best illustrated with an example. Cats are very susceptible to the deleterious effects of many toxic chemicals because they lack several important metabolic pathways. The most important of these pathways is the conjugation of xenobiotics with glucuronic acid. This pathway is particularly important in detoxifying phenols; they are conjugated with glucuronic acid, catalysed by the enzyme glucuronyl transferase, and thus detoxified. The conjugation process forms a very polar molecule which is very well suited to being excreted in the body's water-based excretory systems (e.g. urine). Cats have a very low glucuronyl transferase level (and low levels of several other conjugating enzymes important in the detoxification and excretion of xenobiotics, e.g. sulphohydrolase which adds a sulphate moiety to phenols so detoxifying and making them more polar) and therefore are particularly susceptible to the toxicity of xenobiotics that are metabolised by this route. Perhaps the best example of species-specific toxicity is paracetamol in the cat. Paracetamol (acetominophen; see Figure 6.4) is metabolised in the liver of most animals by its conjugation with either sulphuric acid or glucuronic acid to form very water soluble, non-toxic sulphate and glucuronide conjugates (see Figure 6.4). A very small proportion of the paracetamol is metabolised to an incredibly toxic quinoneimine (see Figure 6.4) which causes cell death. As an aside, paracetamol overdose in humans results in the metabolic generation of more quinoneimine than the cell can detoxify and death results. As the cat is unable (or at best able, but very slowly) to form sulphate and glucuronide conjugates it forces the metabolism of paracetamol down the quinoneimine route. A 'normal' dose of paracetamol to a cat has exactly the same

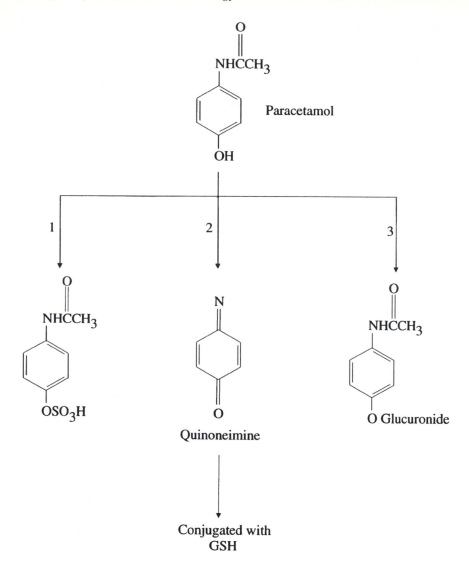

Figure 6.4 The metabolism of paracetamol (acetaminophen) in mammals. Metabolic flux is mainly via pathways 1 and 3. A very small amount of quinoneimine is formed and quickly mopped up by glutathione (GSH). In cats pathways 1 and 3 are minor and therefore very much more highly toxic quinoneimine is formed.

effect as an overdose in humans. This inability to form glucuronide and sulphate conjugates is thought to occur in all members of the cat family. Interestingly, cats have several extra metabolic pathways that most other mammals do not generally utilise; for example they conjugate xenobiotics with the amino acid glutamine. This, however, does not compensate for the lack of sulphate and glucuronide conjugation because different classes of chemicals are detoxified by glutamine conjugation. It is very likely that this metabolic peculiarity has been selected for in the cat family because of a component of the cat's diet that requires this rather unusual metabolic pathway.

This example illustrates the importance of species differences in metabolism in the species-specific toxicity of particular molecules. The domestic cat is not important in any ecosystem (although garden ecosystems suffer from cats killing birds nesting in the hedgerows), but it is very likely indeed that important component species of ecosystems display similar differences in the metabolism of xenobiotics. The greater the prevalance of the particular species in an ecosystem the greater its importance in determining the degradative pathway of a specific environmental contaminant. For example, consider a compost heap as an ecosystem. The most important species in a compost heap are bacteria, fungi and earthworms. All are capable of degrading xenobiotics and each will have different degradative pathways. The overall effect of the compost heap ecosystem upon a xenobiotic will depend upon the proportion of each species present and will represent an 'average' metabolic pathway to which each species contributes.

In order to appreciate the potential for metabolism of xenobiotics in ecosystems it is important to have a 'feel' for the classes of metabolic pathways that might exert their effects upon an environmental contaminant.

6.9.2 *Animal Xenometabolic Pathways*

The majority of studies on the metabolism of xenobiotics has been carried out in mammals because mammals are important as models for humans in assessing the risk of potential medicines. The pathways which are well known in, for example, the rat are very likely to occur in many other animals.

6.9.3 *The Purpose of Xenometabolism*

It is very important not to be teleological when considering the reason why metabolic pathways exist to degrade foreign compounds. We have a tendency to think that xenometabolic pathways have evolved to deal with synthetic environmental contaminants and drugs. This is not true, such pathways evolved long before humans evolved. The pathways almost certainly evolved to allow animals (and to a lesser extent plants) to eat particular foods. For example, if a particular animal ate a plant which contained a chemical component which was toxic to the animal, that animal species would not survive long. If the animal possessed a xenometabolic pathway to allow it to detoxify the plant's toxic components it would be particularly well suited to its environment. This philosophy can be extended to aid evolution and the generation of diversity. An animal that fortuitously possesses a particular metabolic pathway which allows it to add another food plant to its menu is likely to be more successful (in survival and therefore evolutionary terms) than other animals. This diversity will allow the animal to occupy a different ecological niche so reducing the competition from other species (in this case with respect to food requirements). Such an animal would be regarded as successful in evolutionary terms. As species have evolved the more successful ones (and therefore the survivors) have been better equipped to detoxify the potentially toxic components of their food. Some species have taken this to its extreme and have either 'chosen' a food plant that is particularly toxic and so not palatable to most of their competitors or have incorporated toxic components of their food plants into their bodies so that they themselves are not palatable. A good example of this approach to survival is the European cinnabar moth (*Tyria jacobaeae*). The caterpillar of the cinnabar moth lives and feeds on various species of ragwort (the most common being *Senecio jacobaeae*) which contain highly toxic pyrrolizidine alkaloids. The caterpillars are able to detoxify the pyrrolizidines so that they are unaffected by them, but also they accummulate them in their bodies in order to give their predators (e.g. birds) a nasty toxic shock should they decide to include them in their diets. As is often the case, the cinnabar moth caterpillar is very brightly coloured (yellow and black striped) to warn potential consumers of its likely effects. This utilisation of detoxification mechanisms and immunity to toxicity allows the cinnabar moth to occupy a particularly hostile environment (i.e. a toxic plant) and reduces predator pressure. There are many more similar examples which illustrate the importance of xenometabolic pathways in survival.

The metabolism (and therefore detoxification) of environmental contaminants (and drugs) is purely accidental, but fortuitous. The detoxification mechanisms evolved to be non-specific. The more non-specific they were the greater evolutionary advantage they conferred because their non-specificity meant that animals could add different food plants to their diets with relative impunity. This evolved non-specificity meant that they were also able, fortuitously, to metabolise many synthetic molecules (e.g. drugs and environmental contaminants).

The most important enzyme system which fits the criterion of broad substrate specificity is a mixed function oxidase (i.e. an enzyme system that is able to catalyse the oxidation of a broad range of substrates) termed cytochrome P_{450}. Cytochromes form a class of proteins that have an iron atom in their molecular structures; oxidoreduction of iron facilitates their oxidative effects. The 450 refers to the fact that if they are exposed to carbon monoxide they have an optical absorption maximum at 450 nm.

6.9.4 Cytochromes P_{450}

The cytochromes P_{450} form a group of mixed function oxidase enzymes which are present on the endoplasmic reticulum (ER) of many animal cells (in mammals they have a particularly high activity in the liver, but are present in most other organs except the central nervous system). They catalyse a broad array of oxidation reactions, the most important of which are described below.

The products of cytochrome P_{450} activity are generally more polar than the substrate (e.g. an environmental contaminant) and are very well suited, in molecular terms, to further metabolism and excretion by conjugation, for example with glucuronic acid. For this reason cytochrome P_{450}-catalysed reactions are termed phase I reactions and conjugation reactions are termed phase II (see Figure 6.5).

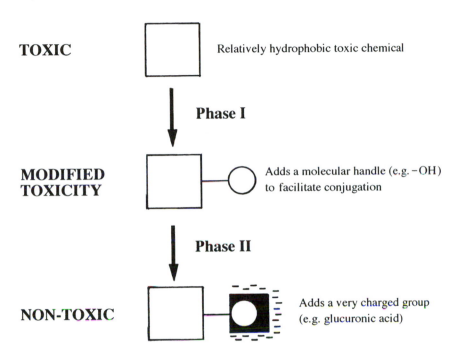

TOXIC Relatively hydrophobic toxic chemical

Phase I

MODIFIED TOXICITY Adds a molecular handle (e.g. −OH) to facilitate conjugation

Phase II

NON-TOXIC Adds a very charged group (e.g. glucuronic acid)

Figure 6.5 Principles of xenobiotic metabolism showing the production of a water-soluble end metabolite to facilitate transport and excretion.

Phase I metabolism of a xenobiotic alters its pharmacological and toxicological profile. This generally results in reduced pharmacological and toxicological activity; however, this is by no means always the case. There are many ostensibly non-toxic molecules which after phase I metabolism are made very toxic indeed (e.g. benzo[a]pyrene, see Figure 6.6); this is termed metabolic activation. It is therefore possible for an environmental contaminant to be made more toxic by the organisms within a particular ecosystem by the action of phase I pathways.

As mentioned above the vast majority of experimental work has been carried out in mammals and therefore the mammalian enzymes are well understood. It is known that many other animals have cytochrome P_{450} activity. Earthworms, mussels and sea urchins all have cytochromes P_{450} which illustrates the broad occurrence of these important detoxification enzymes across the animal world.

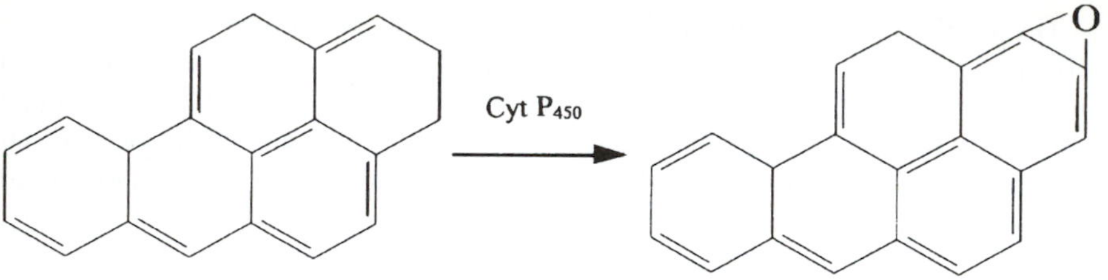

Figure 6.6 Metabolic activation of benzo[a]pyrene to form the highly carcinogenic benzo[a]pyrene epoxide

6.9.5 *Phase I Metabolic Reactions*

This is a complex field and only the most important (and common) reaction types are discussed here. In general, phase I metabolism involves the addition of an oxygen to the substrate. The most common way in which this might be achieved is by the hydroxylation of an aromatic (i.e. containing a benzene ring or a delocalised heterocyclic system such as pyridine) system (aromatic hydroxylation); for example benzene is converted to phenol (see Figure 6.7). As many environmental contaminants are aromatic this is a very important pathway in environmental detoxification. Similarly aliphatic (i.e. straight or branched carbon chains, for example $CH_2CH_2CH_2CH_2CH_2CH_3$ [hexyl]) hydrocarbons are hydroxylated to form alcohols (see Figure 6.7). Some of the cytochrome P_{450}-catalysed reactions do not, on first observation, appear to be oxidative; for example dealkylation reactions. In these reactions the methyl group of toluene can be removed by cytochrome P_{450} (although this tends, in most species, to be a minor route of metabolism; see Figure 6.8). The mechanism of dealkylation involves the initial hydroxylation of the aliphatic group (see Figure 6.8) followed by its removal as an aldehyde. Another important oxidative metabolic reaction catalysed by the cytochromes P_{450} is oxidative deamination (see Figure 6.9). This pathway allows the removal of an

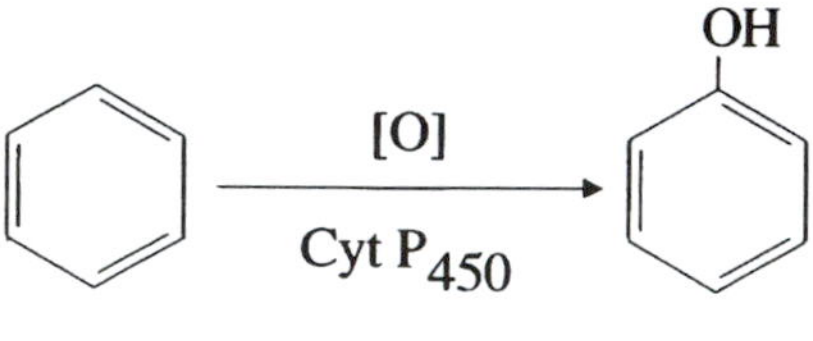

$$CH_3CH_2CH_2CH_2CH_2CH_3 \xrightarrow[\text{Cyt } P_{450}]{[O]} CH_3CH_2CH_2CH_2CH_2CH_2OH$$

Figure 6.7 Cytochrome P_{450}-catalysed hydroxylation of benzene (above) to form phenol and hexane (below) to form hexan-1-ol.

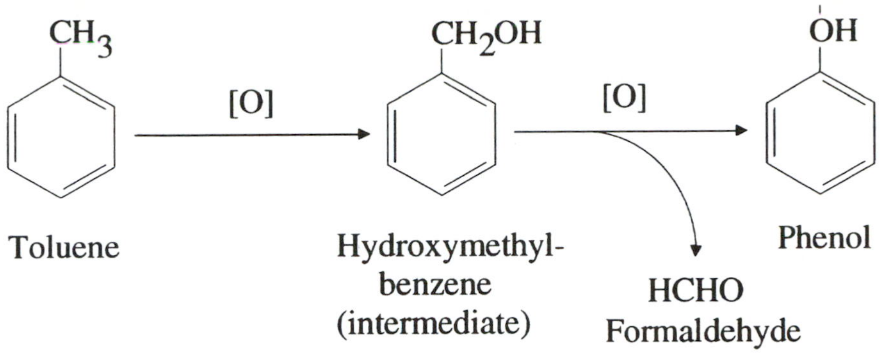

Figure 6.8 Cytochrome P_{450}-catalysed oxidative demethylation of toluene to form phenol.

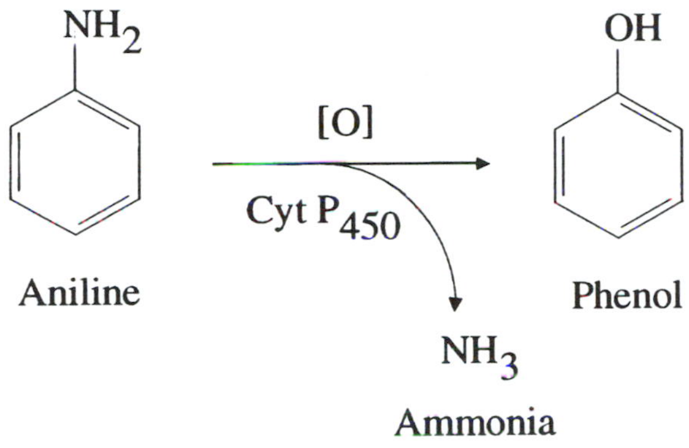

Figure 6.9 Cytochrome P_{450}-catalysed oxidative deamination of aniline to form phenol.

amino group from either an aromatic or an aliphatic system. For example, aniline might be metabolised to phenol via this route.

It is clear from this diverse array of metabolic pathways that significant modifications can be made to a molecule that enters an ecosystem which has animal components. The changes to the molecule will change its toxicity profile (which might involve an increase in toxicity) but, much more importantly, will result in an increase in polarity. This increase in polarity is very important indeed in the overall fate of the molecule in the environment because the greater the polarity of the molecule the lower its persistence and therefore the molecule will have a lower impact in the long term.

To illustrate the potential impact of cytochrome P_{450}-catalysed biotransformations upon an environmental contaminant a hypothetical example (1-chloro, 2-ethyl, 4-aminobenzene; see Figure 6.10) will be considered. From this single molecule it is possible to produce at least 11 metabolites (and this is a very

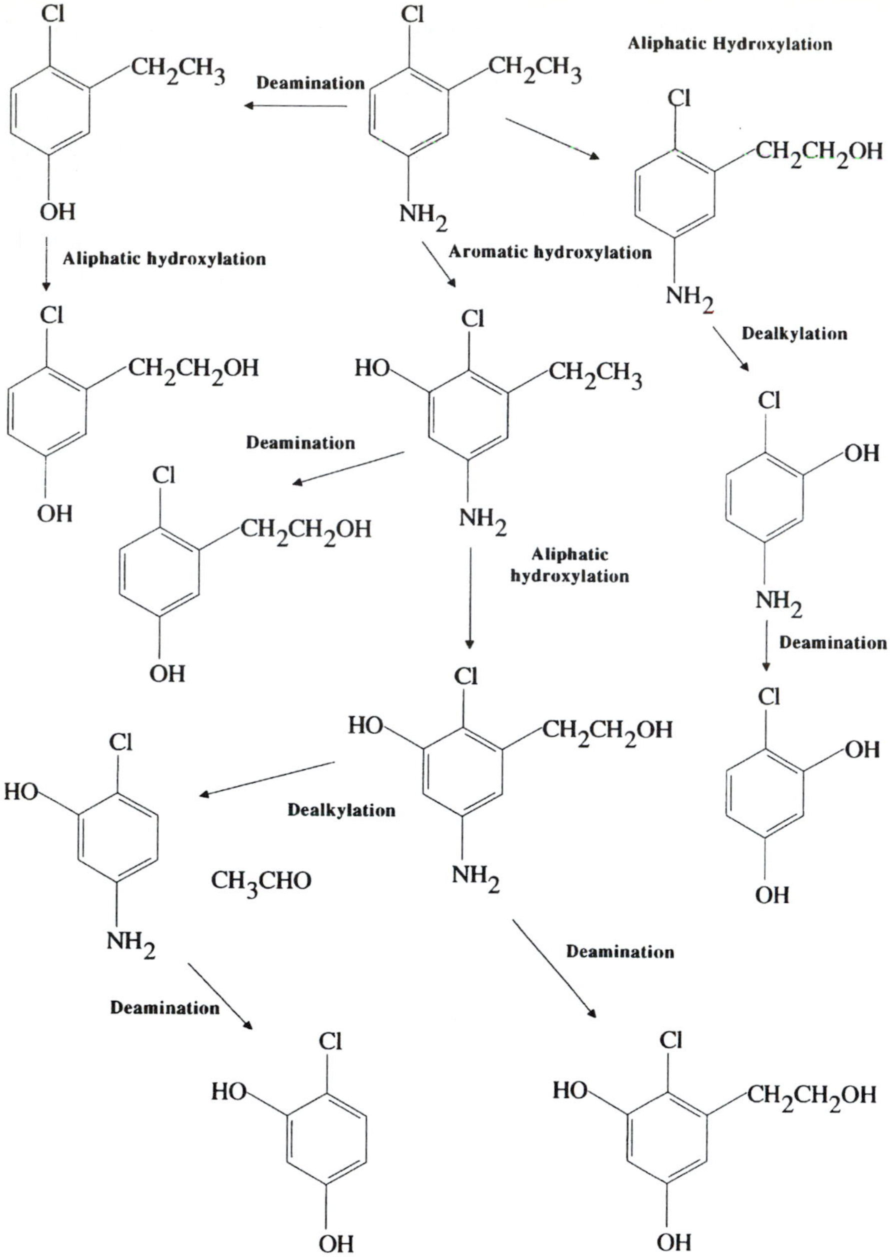

Figure 6.10 Possible cytochrome P_{450}-catalysed biotransformations of a hypothetical environmental contaminant (1-chloro, 2-ethyl, 4-aminobenzene) showing the array of hypothetical metabolites.

conservative estimate). It is, perhaps, unlikely that quite so many metabolites would be produced in a 'real' ecosystem because one or two pathways would be likely to predominate. This is best illustrated by a real example.

The fate of DDT in the environment illustrates that a limited number of metabolic pathways tend to predominate (see Figure 6.11). Its main environmental metabolite is *p,p'*-dichlorophenyldichloroethene (*p,p'*-DDE), although a small amount of the intermediate metabolite *p,p'*-dichlorodiphenyldichloroethane (*p,p'*-DDD) also persists. Looking at the molecular structure of DDT (see Figure 6.11) one might predict that phase I hydroxylation would occur; however, it does not. The purpose of using this example is to add caution to predicting the fate of environmental contaminants.

An important part of the fate equation is where in the ecosystem the molecule is present. Is it accessible to degradative enzymes? If the answer is no, then the enzymes simply cannot act upon the molecule. DDT probably falls into this

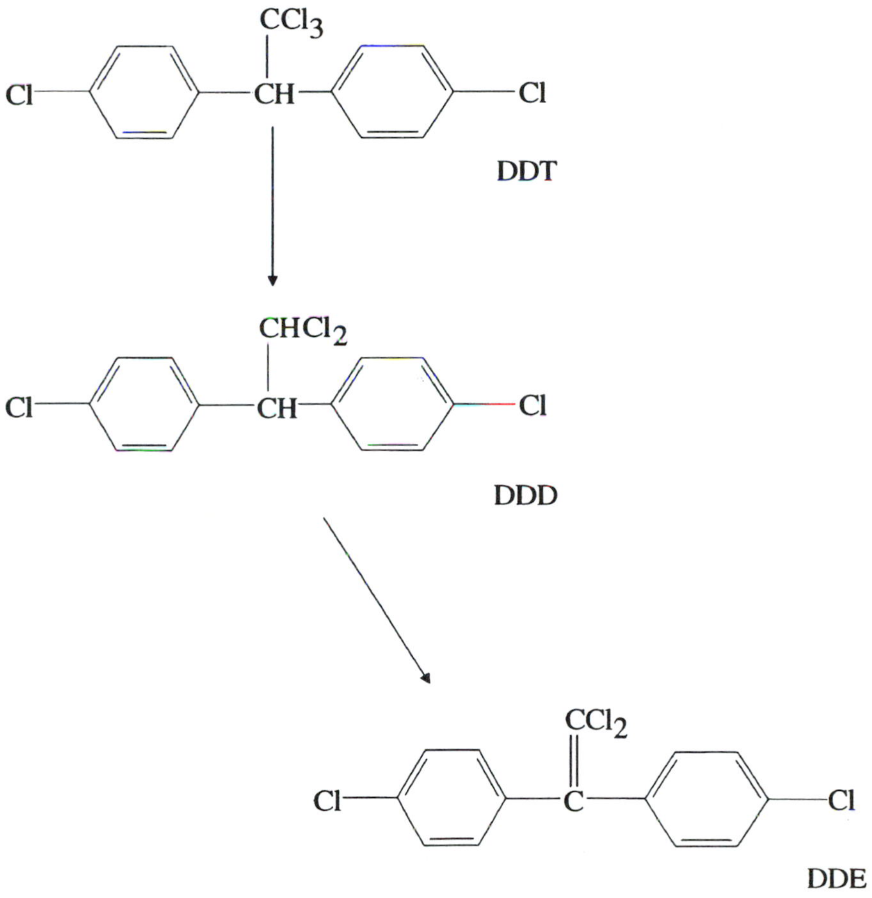

Figure 6.11 Biotransformation of DDT showing the formation of DDD and DDE.

category because it is so lipophilic that it is taken up immediately by cells and goes no further than their cell membranes. Cytochromes P_{450} never have access to the molecule and so oxidation reactions do not readily occur.

To further illustrate the environmental fate hypothesis we will consider a very much more water-soluble contaminant, namely the OP pesticide Diazinon. It is very much more quickly degraded than DDT (Diazinon $t_{1/2}$ in soil = 8 days, DDT $t_{1/2}$ = approx. 12 years; see Figure 6.12).

6.9.6 *Phase II Metabolic Reactions*

Molecules with appropriate handles (see Figure 6.5) to facilitate conjugation (e.g. hydroxyl groups) can go straight into phase II metabolism. Those without need (i.e. because they already have, for example, OH) pass directly to phase II.

The main outcome of phase II metabolism is a high molecular weight, very water-soluble and non-toxic molecule. The molecules which cells use to conjugate are very varied indeed and what is more they vary greatly with species. Those most commonly encountered are glucuronic acid (an oxidation product of glucose), sulphuric acid, the amino acid glycine and the ubiquitous cell-protecting molecule glutathione.

6.9.6.1 *Glucuronic Acid Conjugation*

Any molecule with either a hydroxyl group or a carboxyl group can be glucuronidated (see Figure 6.13). Hydroxyl compounds form glucuronyl ethers and carboxyls form glucuronyl esters. The latter are far less stable than the former. The glucuronidation reaction is catalysed by a group of deeply membrane-embedded enzymes, the glucuronyl transferases. One could imagine the enzyme sitting in the endoplasmic reticulum in the vicinity of the cytochromes P_{450} picking up the hydroxylated xenobiotic metabolites and immediately forming glucuronide conjugates before they are safely transported out of the cell.

In the environmental situation it is very likely that a large proportion of glucuronide conjugates released from animals into ecosystems will be degraded to release the hydroxyl or carboxyl xenobiotic because of the presence of bacterial β-glucuronidase (an enzyme which breaks down glucuronide conjugates). This might increase the toxicity again and restart the process.

6.9.6.2 *Sulphuric Acid Conjugation*

This process is similar to glucuronidation except that the sulphotransferases are cytoplasmic. They catalyse the transfer of the sulphate moiety from a metabolically activated form onto a hydroxyl group to form an organic sulphate (see Figure 6.14). The sulphate conjugate is very water soluble and non-toxic and can be transported out of the cell safely. As with the glucuronides, there are bacterial enzymes present in many environmental systems (sulphohydrolases)

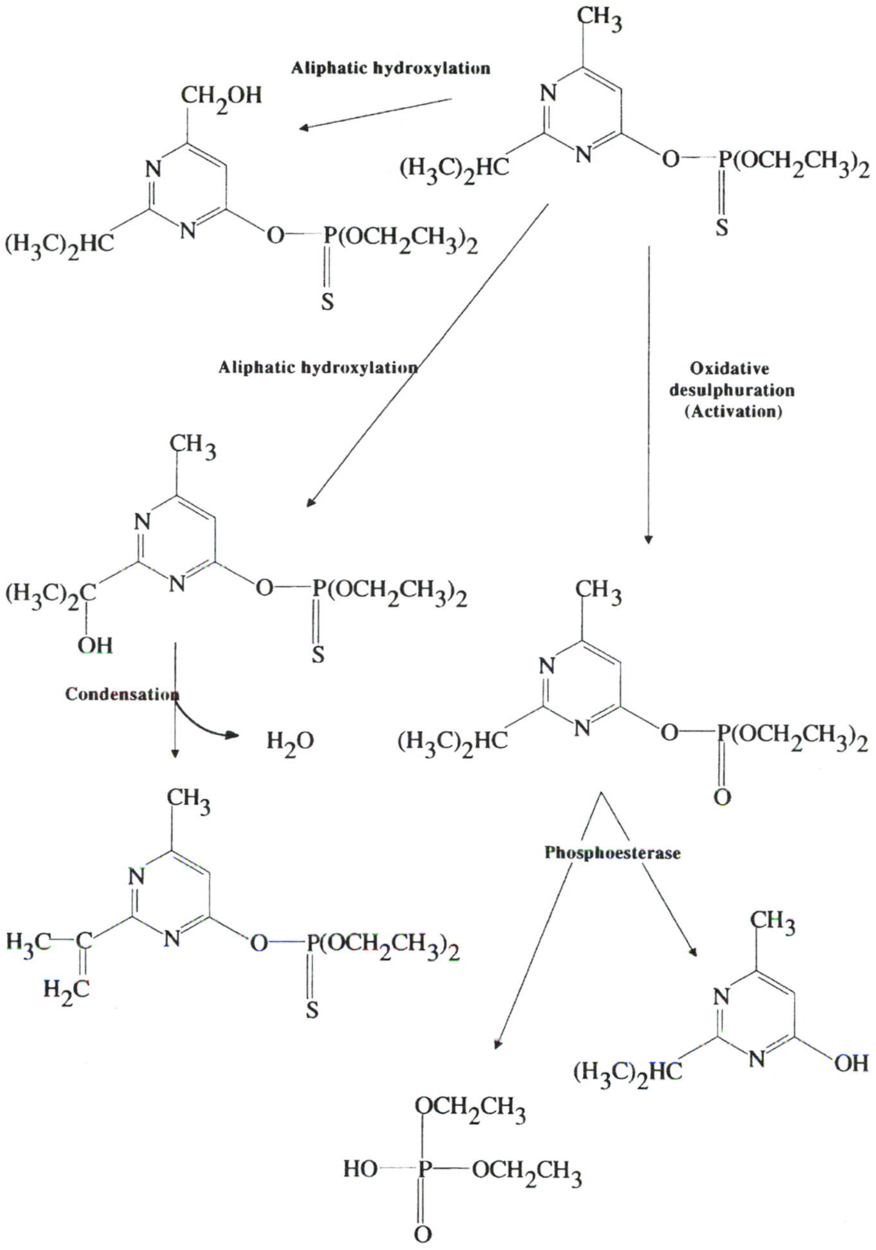

Figure 6.12 Part of environmental degradative pathway of the OP Diazinon showing its activation to the OXO compound.

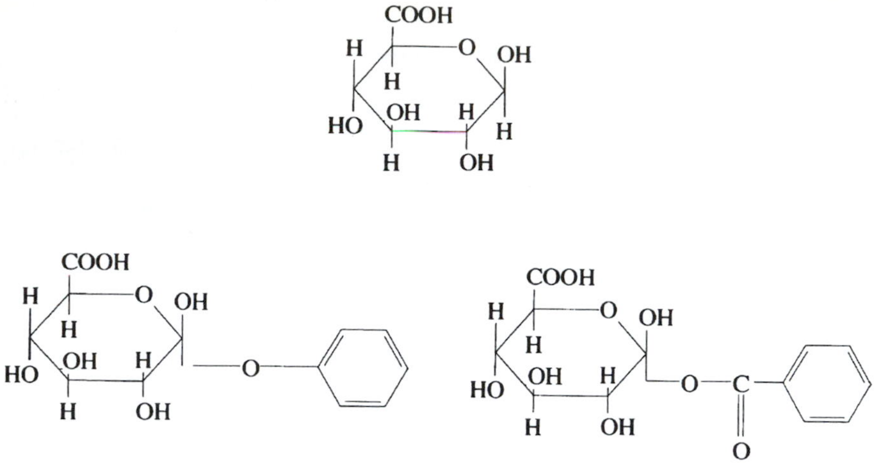

Phenylglucuronide (an ether)   Benzylglucuronide (an ester)

Figure 6.13 Formation of ether and ester glucuronides of benzoic acid and phenol. Glucuronic acid is shown above the two conjugates.

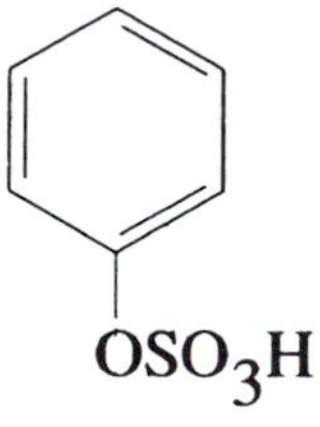

OSO_3H

Figure 6.14 Phenyl sulphate, the sulphate conjugate of phenol.

which break down sulphate conjugates to release the original molecule or phase I metabolite.

6.9.6.3 *Amino Acid Conjugation*

By far the most common amino acid involved in phase II reactions is glycine, although, as discussed above, cats utilise several amino acids including glutamine. Glycine conjugates are formed by a complex series of metabolic reactions which results in the formation of an amide (peptide) bond between the carboxyl group of a xenobiotic and the amino group of the amino acid (see Figure 6.15). Amino acid conjugates are excreted from the animal and are more stable than sulphate and/or glucuronide conjugates in the environment.

138

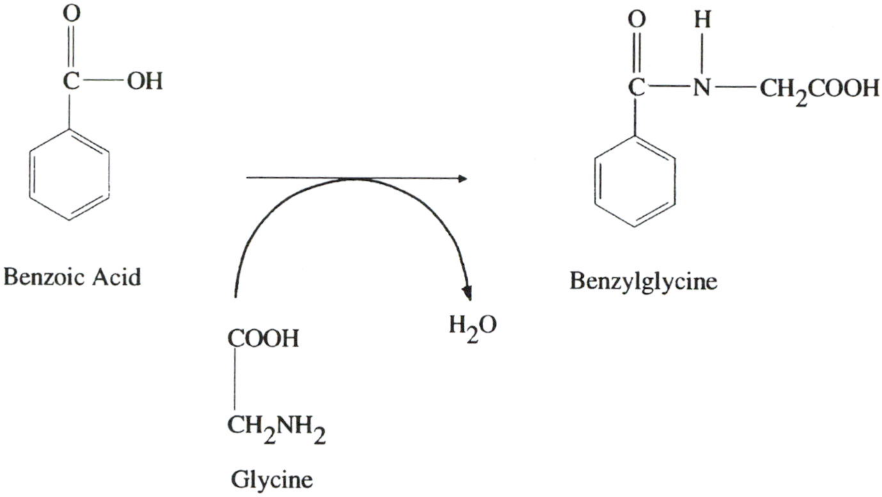

Benzoic Acid

Benzylglycine

COOH

H₂O

$$COOH$$

$$CH_2NH_2$$

Glycine

Figure 6.15 Conjugation of benzoic acid with glycine to form benzylglycine.

6.9.6.4 *Glutathione Conjugation*

Glutathione (GSH) is a tripeptide consisting of glutamic acid, cysteine and glycine residues (see Figure 6.16). It has a free sulphydryl (from the glycine residue) which is important in its chemical reactions. GSH is present in a great many animal cells as a general cell-protecting molecule. It reacts spontaneously with many reactive xenobiotics or reactive species (e.g. free radicals) formed by normal metabolic processes (e.g. superoxide formed by xanthine-oxidase-catalysed metabolism of xanthine to form uric acid). In addition to its spontaneous reactions there is a group of enzymes, the glutathione transferases, which catalyse the conjugation of GSH with a wide range of xenobiotics (e.g. chlorinated aromatic compounds; see Figure 6.17).

GSH conjugates are metabolised further in most animal cells by sequential removal of the component amino acids of GSH to leave the cysteine conjugate

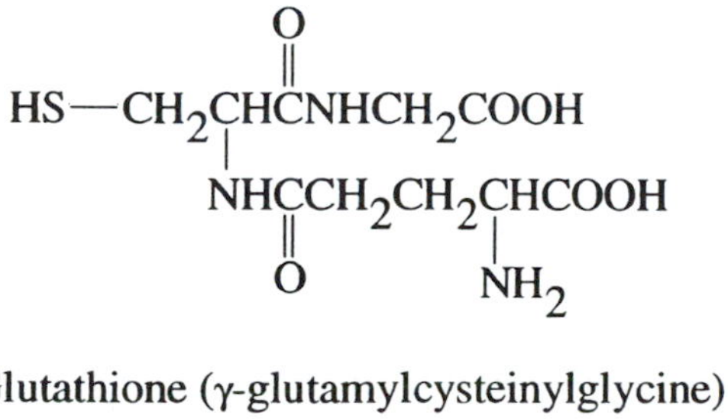

Glutathione (γ-glutamylcysteinylglycine)

Figure 6.16 Molecular structure of the tripeptide glutathione.

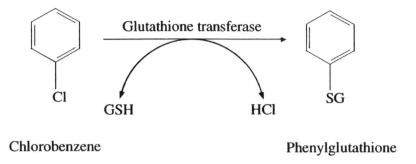

Chlorobenzene   Phenylglutathione

Figure 6.17   Conjugation of chlorobenzene with glutathione (GSH).

which is *N*-acetylated to form a mercapturic acid (captan signifies sulphur content and uric pertains to its urinary excretion). It is very likely that many chlorinated aromatic environmental contaminants are at least in part released as mercapturic acids into ecosystems by the animals within the system.

Phase I and phase II metabolism are only facets of the potential battery of molecular modifications that an environmental contaminant might undergo. They are ostensibly carried out by animal cells and perhaps are a minor contribution to the fate and behaviour of a xenobiotic in an ecosystem.

It is very difficult (and misleading) to generalise as to which might be the most important degradative pathway for xenobiotics in the environment, but it is certain that microorganisms play a very important role, if only because of their enormous numbers in all ecosystems. Another important point about microorganisms is that they are able to carry out molecular degradations which mammals simply are unable to effect. For example, bacteria are very good at breaking aromatic rings into fragments. This is clearly a very important reaction in the total degradation of an environmental contaminant.

6.10   Bacterial Degradation of Environmental Contaminants

Surprisingly there is not a great deal known about xenobiotic degradative pathways in microorganisms which are relevant to environmental toxicology. Much work has been carried out on the gut microfloral metabolism of drugs and the findings of such studies can be extrapolated to the environment. It is important to remember that there are very many species of microbes in the biosphere and that they are likely to exhibit species differences in the metabolism of foreign compounds.

In this section we generalise and only cover the types of, rather than specific, pathways catalysed by bacterial enzymes.

Perhaps the most important reactions are ring-breaking or -scission reactions. Bacteria are able to break bonds in heterocyclic rings, generally between the heterocyclic atom (e.g. O in pyrans) and its adjacent carbon. The product, which might be a hydroxyl, aldehyde or carboxylic acid, might form the substrate for animal cell-mediated reactions and is likely to initiate the biodegradation of classes of molecules that animal cells would otherwise be unable to bring about.

In addition to the ring-scission reactions bacteria can perform oxidative reactions, for example they will oxidise hydroxylaliphatic compounds to carboxylic acids. They can decarboxylate and release carbon dioxide, so an aliphatic alcohol might be oxidised to its corresponding carboxylic acid and the carboxyl group be removed so reducing the carbon chain length by one. They are able to carry out β-oxidation reactions whereby long aliphatic chains (e.g. fatty acids) are sequentially diminished in length by two carbon units at a time. They can remove amino groups from both aromatic and aliphatic situations and release ammonia (hence the smell of ammonia when you dig into a compost heap or drive past a sewage treatment works). Bacteria are very versatile indeed and are able to degrade many molecules completely.

The versatility of bacteria in the context of the degradation of environmental contaminants is illustrated well by the use of activated sludge systems in sewage treatment works. We have seen a very graphic illustration of the efficiency of bacterial degradation in a sewage treatment works in Yorkshire in the northeast of England. The works treats water from a broad range of industries including several dye manufacturers. Occasionally the effluent received at the sewage treatment works is brightly coloured but on passing this through the activated sludge system a colourless liquid is produced. The colour change is likely to be due to the bacteria's heterocyclic ring-scission capabilities because many dyes have nitrogen atoms in an aromatic ring structure to give them their characteristic colour.

Bacteria are without doubt the most important members of an ecosystem in terms of biodegradation. Without them the earth would quickly become saturated with the enormous amount of xenobiotics that we release every day.

Further Reading

Briggs, G.E., 1990, Predicting the behaviour of pesticides in soil from their physical and chemical properties, *Philosophical Transactions of the Royal Society of London*, **329**, 375–82.

Hemond, H.F., 1994, *Chemical Fate and Transport in the Environment*, San Diego: Academic Press.

Klaassen, C.D., 1996, *Casarett and Doull's Toxicology — the Basic Science of Poisons*, 5th Edn, UNIT 2 — The Disposition of Toxicants, New York: McGraw-Hill.

7

Radioactivity and the Environment

The controversy surrounding the use of radioisotopes as nuclear fuels and the impact of radioactive pollutants upon the environment are discussed.

The concepts of radioactive decay and background radioactivity are introduced. Examples of sources of pollutants and their impact are discussed:

- Nuclear power stations
- Chernobyl
- Leukaemia clusters
- Radon
- Nuclear war

7.1 Introduction to the Issues

Radioactivity and its impact upon the environment and humans is probably the most exaggerated problem of our time. The word radioactivity conjures up all sorts of horror in many people's minds. The horrors (i.e. hazards, see Chapter 1) are all true, but the risk of their happening is very much less than the press and some environmental groups would have us believe. For example, the hazards associated with the intake of α-emitting isotopes (e.g. some plutonium nuclides) are great: if inhaled they are capable of causing lung cancer and therefore death. The chance of this happening (i.e. the risk) is very low when the isotopes are used in the generation of electricity. The risk assessment therefore points to low risk. If we introduce the benefit into the equation the result is even more favourable. We live in a world where our traditional energy sources (fossil fuels: coal, oil, gas) are rapidly running out. We have to develop a viable alternative quickly (i.e. on a century scale). Nuclear power is an excellent alternative. It uses a sustainable (or even producible) fuel (plutonium), is clean (produces no toxic exhaust gases) and can be phased into our current power networks with minimum disruption. The risk:benefit equation is therefore very far on the side of benefit.

By now a good many readers will be incensed, partly because they have been persuaded by what they have read elsewhere and partly because a major part of the risk argument has been missed: if a major accident involving the generation of nuclear power does occur it could be a disaster of major proportions. The question is, can we accept this possibility? The answer is extremely controversial and is related to how desperate we are for electricity. At the moment the problem is a hundred or so years down the line, it does not affect us directly and therefore we as a community are not prepared to accept the risk, however small. At one time rational thinking began to persuade many people that we must think about our descendents and then the Chernobyl disaster occurred and this incredibly remote risk became reality. Nuclear power is now guaranteed to introduce a frown to the faces of most environmentally conscious people.

It is interesting that the degree of acceptance or opposition to nuclear power is, to some extent, a national trait. The French are rational and accept the inevitability of nuclear power and have a good programme to introduce and research the nuclear generation of electricity. The Americans are far more sceptical, but then they remember a near nuclear disaster at the Three Mile Island power plant in 1979 and who can blame them for requiring more assurance before they embrace the need to generate electricity by nuclear means. The UK has swung from a cautious acceptance to absolute opposition largely because of a popular misconception of the risks (i.e. perceived risk > actual risk). The Scandinavians are anti-nuclear power. This is generally thought to be due to their environmentally aware nature; however, it must be influenced by the fact that they have a hydroelectric resource and therefore do not need an alternative power source. The problem, of course, is that differences in international opinion are irrelevant in the context of environmental (and human exposure) risk because if an accident occurs it is the weather patterns that decide who is exposed and where the contamination occurs rather than the boundaries between countries. This is illustrated very well by the Chernobyl disaster when Scandinavia was contaminated because of the wind patterns immediately after the disaster. It is also interesting to note that several French nuclear power stations are situated on their north coast and that they are closer to major UK cities than to many major French cities. Nuclear power is a hot political issue and will remain so for many years to come. Perhaps opinion will change when fossil fuels run out.

7.2 Radioactive Decay

It is important that before embarking upon a discussion of the environmental impact of radioactivity that the fundamentals of radioactive decay are appreciated. There are three basic means by which a radionuclide can decay (see Table 7.1); (1) by loss of a fragment of its nucleus identical in form to the helium nucleus (i.e. 2 protons and 2 neutrons with no attendant electrons, hence He^{2+}) – α-emission, (2) by loss of an electron – β-emission, or (3) by loss of electromagnetic energy – γ-emission. It is probable that the daughter atom is also radioactive and unstable and will decay further until a stable atom is formed. This is termed a decay chain (see Figure 7.1). It is very likely indeed that a decay chain will involve a mixture of

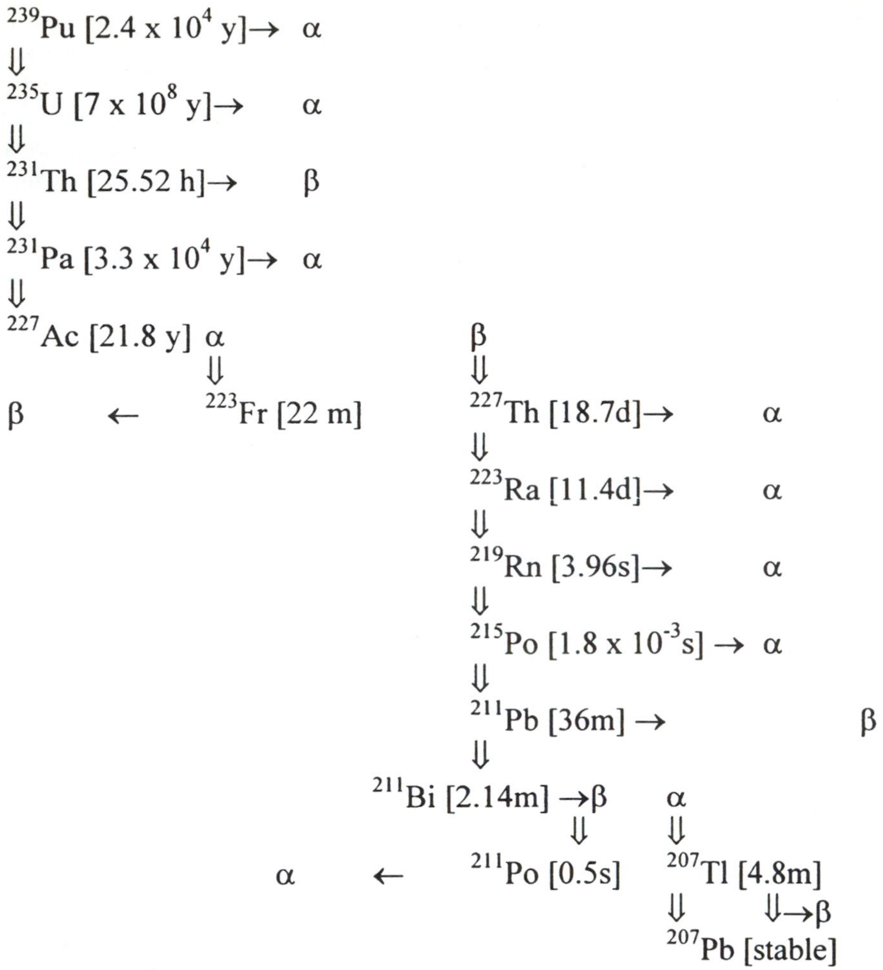

^{239}Pu [2.4 x 10^4 y]→ α
⇓
^{235}U [7 x 10^8 y]→ α
⇓
^{231}Th [25.52 h]→ β
⇓
^{231}Pa [3.3 x 10^4 y]→ α
⇓
^{227}Ac [21.8 y] α   β
        ⇓   ⇓
β   ← ^{223}Fr [22 m] ^{227}Th [18.7d]→   α
                                       ⇓
                                   ^{223}Ra [11.4d]→   α
                                       ⇓
                                   ^{219}Rn [3.96s]→   α
                                       ⇓
                                   ^{215}Po [1.8 x 10^{-3}s] → α
                                       ⇓
                                   ^{211}Pb [36m] →   β
                                       ⇓
                      ^{211}Bi [2.14m] →β   α
                              ⇓   ⇓
α   ← ^{211}Po [0.5s] ^{207}Tl [4.8m]
                                       ⇓   ⇓→β
                                   ^{207}Pb [stable]

Figure 7.1 Decay chain for ^{239}Pu showing that the daughter nuclides are also radioactive. The decay eventually leads to the formation of a stable lead. γ-Decay occurs at almost every step in the chain, but is not shown.

radioactive decay types; indeed, it is also probable that an individual isotopic decay will involve more than one emission type.

The energy of radioactive decay is measured in electron volts, usually millions of electron volts (MeV). α-Particles have energies in the region of 3–6 MeV, β-particles in the range 0.005–3 MeV and γ- rays 0.1–2 MeV.

Radioactivity is measured in bequerels (previously the unit of radioactive decay was the curie), named after the French physicist who was the first to recognise the existence of radioactivity. One bequerel (Bq) is 1 disintegration per second (1 curie (Ci) = 3.7 × 10^{10} disintegrations per second). To express how radioactive a substance is, specific activity (SA) is used:

Table 7.1 Some characteristics of the three basic forms of radioactive decay

Type of decay	Form	Characteristics	Physical barrier
Alpha — α	He^{2+}	High energy Low penetration High mass Will not penetrate skin	Paper, 5 cm air
Beta — β	e^-	Moderate energy Very variable energy Medium penetration Will penetrate skin	Perspex, metal film

Gamma — γ	Electromagnetic	Moderate–high spectrum
		    of energy
		High penetration
		Will pass through body
			1 cm lead, thick
			    concrete

$$SA = \frac{\text{Radioactivity of material (Bq)}}{\text{Amount of material (moles or g)}}$$

$$= \text{Bq mol}^{-1} \text{ or Bq g}^{-1}$$

The rate of radioactive decay is measured as half-life ($t_{1/2}$; sometimes referred to as half period). This is the time taken for 50 per cent of a particular isotope to decay to its daughter. This time is extremely variable between isotopes, for example the $t_{1/2}$ of $^{14}C = 5730$ years whereas the $t_{1/2}$ of $^{13}N = 9.96$ minutes. A long half-life means that the chance of a radioactive particle being emitted at a particular time is low, but that the isotope will persist for many years in the environment or body. In general, environmental toxicologists are more worried about long half-life isotopes than short half-life isotopes. We return to these issues when the Chernobyl disaster is discussed in detail later in this chapter. In essence, there were several isotopes released from Chernobyl, some of which had a short half-life (e.g. ^{131}I, $t_{1/2} = 8$ days). They disappeared from the environment quickly and therefore were only a problem to people and animals who received a high initial dose because they lived in the immediate environs of the site (iodine is concentrated in the thyroid gland and the incidence of thyroid cancer has increased in people in the Chernobyl region). On the other hand, there were also relatively long-lived isotopes (e.g. ^{137}Cs, $t_{1/2} = 30$ years) which are going to be with us and contaminate the food chain for at least a century (after 90 years there will be 12.5 per cent of the original activity of ^{137}Cs in the environment). The environment and humans will therefore receive a long-term, relatively low dose exposure to their emissions. It is likely that the general incidence of cancer will rise (marginally) as a result of exposure to these persistent isotopes. This is, however, extremely difficult to measure and even more difficult to ascribe to a particular exposure because there are so many coincident variables.

7.2.1 Background/Natural Radioactivity

Most elements comprise a mixture of isotopes, a proportion of which are unstable and therefore radioactive. In general, the radioactive component of the elements which make up the earth is a minute proportion of the total mass of the element. For example, potassium has an atomic weight of 39.0983 daltons. This fractional atomic weight is due to potassium being composed of a mixture of isotopes in different proportions; the atomic weight of potassium is a weighted average of its component isotopes (see Table 7.2).

Anything which contains an element that has a radioactive component in its isotopic mixture is in turn radioactive. For example, potassium is ubiquitous in nature, it is an important electrolyte (as KCl) in cells and therefore every plant and animal is being internally bombarded with β-particles and γ-rays emitted from the ^{40}K component of the potassium within their cellular structure. In terms of risk this is extremely small and unavoidable, but it must contribute towards the background incidence of cancer. Some plants (and perhaps animals too) selectively take up and concentrate potassium; such plants are therefore more radioactive than other plants. If such a plant were used as food by animals and/or humans its consumer would be ingesting ^{40}K over and above the 'normal' activity of other foodstuffs. There are several foods which fall into this category; Brazil nuts and coffee are potassium rich and therefore each time you drink a cup of coffee or eat a Brazil nut you are ingesting more radioactivity than with most other foods. This is just one example of background or natural radioactivity.

Background radioactivity is a useful concept because it provides a good bench mark against which to measure exposure to 'unnatural' isotopes. For example, one could measure ingestion of ^{37}Cs from the Chernobyl disaster as being equivalent to drinking a certain number of cups of coffee per year. This is an excellent way of setting radioactive exposure in perspective.

Another important aspect of natural radioactivity is external exposure. In this case an individual is exposed to radioactive emissions from the outside of their bodies, rather than ingesting them with their food and their cells being directly exposed by the isotope within. There are several elements which are important in this respect, perhaps the most important is uranium. Uranium is a component of several rock types (e.g. granite) and, therefore, because uranium is radioactive and produces a series of radioactive daughters (see Table 7.3), granite emits β-particles and γ-rays. If you are a keen walker and walk over regions rich in granite you will be exposed to more radioactivity than if you choose (for

Table 7.2 Isotopic composition of natural potassium

Isotope	Proportion (%)	Emission	$t_{1/2}$
^{39}K	93.22	Stable	
^{40}K	1.2×10^{-2}	β and γ	1.3×10^{9}
^{41}K	6.77	Stable	

Table 7.3 Uranium decay chain

Isotope		Emission	$t_{1/2}$
^{238}U	→	α, γ	4.5×10^9 years
⇓			
^{234}Th	→	β	24 days
⇓			
^{234}Pa^a	→	β, γ	1.17 min
⇓			
^{234}U	→	α, γ	2.45×10^5 years
⇓			
^{230}Th	→	α, γ	7.54×10^4 years
⇓			
^{226}Ra	→	α, γ	1.6×10^3 years
⇓			
^{222}Rn	→	α, γ	3.8 days
⇓			
^{218}Po	→	α	3.1 min
⇓			
^{214}Pb	→	β, γ	26.8 min
⇓			
^{214}Bi	→	β, γ	19.9 min
⇓			
^{214}Po	→	α, γ	164 microseconds
⇓			
^{210}Pb	→	β, γ	22.3 years
⇓			
^{210}Bi	→	β	5 days
⇓			
^{210}Po	→	α, γ	138.4 days
⇓			
^{206}Pb (stable)			

^{a} Metastable.

example) a limestone region. More importantly though, if you live in a granite house you will be exposed for long periods of time to the emissions from the uranium and its radioactive daughters within the fabric of your home. Cornwall, a south-western county of England, is particularly rich in uranium and it has been calculated that a day lying on a Cornish beach far exceeds a lifetime's exposure to plutonium emitted from nuclear power stations. Again, this puts exposure to radioactivity in perspective.

7.2.2 *Plutonium*

Plutonium (Pu) is an element which stikes terror into many an environmental activist's heart. It is common to see graffiti on road bridges and walls proclaiming 'plutonium-free zone'. Perhaps the reason for this animosity towards a potentially very useful element is its use in nuclear weapons; it certainly is not due to its environmental contamination because this is extremely low. Plutonium is extremely toxic, mainly because of its α-emitting property but it is possible that the metal *per se* has intrinsic toxicity.

Plutonium has been known to humans for only 50 years. It is found in nature, but most of the plutonium present on earth today is synthetic. This is another reason for our concern: we are manufacturing a very toxic element. Until quite recently it was thought that plutonium was entirely synthetic in origin; however, the existence of natural reactors from many millennia ago has been discovered. These natural reactors were areas of high concentrations of uranium in the earth's crust; the conditions were such that plutonium was generated in exactly the same way that nuclear reactors of today generate it. ^{239}Pu was formed by these natural reactors and because it was formed many half-lives ago ($t_{1/2} = 24\,065$ years) most has now decayed. There is some natural ^{244}Pu on earth today which originated from the formation of the solar system. Plutonium is therefore not entirely synthetic but in excess of 99.999 per cent of the element on earth today is manufactured.

There are many isotopes of plutonium (see Table 7.4) and all are radioactive. Generally when plutonium is talked about it is ^{239}Pu which is being discussed because this is the isotope which is important in power generation and the manufacture of nuclear weapons.

Table 7.4 Isotopes of plutonium showing their radioactive emissions and daughters

Isotope half-life (years)	Main emissions	Daughter half-life (years)	Main emissions
Plutonium 237 0.124	Photon—115 keV	Neptunium 237 2 140 000	α — 4.8 MeV
Plutonium 238[a] 87.74	α — 5.5 MeV	Uranium 234 244 500	α — 4.8 MeV
Plutonium 239[a] 24 065	α — 5.2 MeV	Uranium 235 703 800 000	α — 4.4 MeV
Plutonium 240[a] 6537	α — 5.2 MeV	Uranium 236 23 415 000	α — 4.6 MeV
Plutonium 241[a] 14.4	β — Ave, 5.2 keV	Americium 241 432.2	α — 5.5 MeV
Plutonium 242[a] 3 763 000	α — 4.9 MeV	Uranium 238 4 468 000 000	α — 4.2 MeV
Plutonium 244 82 600 000	α — 4.6 MeV	Irrelevant as parent half-life is so long!	

[a] Present in spent fuel in significant quantities and hence at the reprocessing stage.

7.2.2.1 *Plutonium Toxicity*

There has been much debate amongst toxicologists on whether plutonium is toxic solely due to its α-emissions or because of some intrinsic properties of the metal itself. This will, of course, never be resolved because all of the isotopes of plutonium are radioactive and therefore it is not possible to carry out a toxicity study with a non-radioactive form to investigate non-radionuclide effects. ^{241}Pu, however, is a β-emitter and therefore it is possible that the importance of the α-particle in its mechanism of toxicity might one day be investigated. This debate is purely academic; plutonium is exceedingly toxic by whatever means.

It is thought that plutonium's mechanism of toxicity relies upon its localisation in the cell in the vicinity of the nucleus; the α-particle emitted from the plutonium can then directly bombard (for example) DNA and result in base changes which in turn result in either cell death or transformation. The latter leads to the development of a tumour. It is important that the plutonium is actually within the cell because the penetration distance of the highly energetic (5.2 MeV from ^{239}Pu) α-particles is extremely short (40 μm in soft tissues). If the plutonium were outside the cell it is likely that the α-particles would not reach the nucleus. The extremely long half-life (24 065 years for ^{239}Pu) of plutonium means that deposits of the element within a cell would bombard the cell for a very long time indeed; undoubtably for far in excess of the cell's life expectancy.

Another means by which plutonium can damage cells is by bombarding them from the outside and disrupting the cell membrane. This results in cell death and initiates cell proliferation which might result in tumour development by a non-genotoxic mechanism. This mechanism of toxicity is particularly important when plutonium is inhaled because it can reside in the lungs for many years. During this time the cells of the lung in the immediate vicinity of the deposit are bombarded by highly energetic α-particles.

There is good evidence from animal studies that the major toxic outcome of plutonium exposure is cancer. Studies in dogs have shown that injection of plutonium salts into a vein or into the skin results in bone or, less frequently, liver cancer. Bone and liver are the cancer targets because these are the two tissues to which plutonium is transported when administered. Plutonium is transported on the iron carrier protein transferrin in the circulatory system. Specific exchange mechanisms occur in bone which means that the plutonium is released from the carrier and deposited on the surface structure of the bone where it can stay for many years. During this time it bombards the active osteocytes on the bone with α-particles resulting in their transformation into cancer cells. One route of elimination of plutonium from the body is in bile, suggesting that plutonium is taken up in the Kupffer cells of the liver (possibly while bound to transferrin). Liver cancer results from exposure during the period of time that the plutonium resides in the liver. Bone cancer is the most common tumour associated with plutonium administration to animals; liver cancer is far less common. One reason for this differential cancer incidence is because once the plutonium is deposited on bone it remains there for very many years (biological $t_{1/2} \geq 12$ years), whereas its residence time in the liver is very much shorter because of its relatively rapid elimination in the bile. The cumulative radiation dose in bone is therefore significantly greater than in liver hence the greater cancer incidence in bone. There is no evidence that bone or liver cancer results from human exposure to plutonium. If plutonium is inhaled by dogs, primary lung cancer develops due to deposits of plutonium remaining in the lung for long periods of time.

7.2.2.2 *Ecotoxicity of Plutonium*

It is very likely that ingestion of plutonium by any animal may result in the development of tumours. This is likely to be insignificant in the context of the survival of ecosystems as a whole. Plutonium is genotoxic and therefore if sex cells are contaminated it might result in mutations that could affect future generations. This is much more likely to have environmentally significant effects. Very little work has been carried out in this area and therefore it is only possible to speculate, but as the raw material of evolution is mutation, increasing the frequency of mutations might increase the speed of evolution.

7.3   Power Stations

There are four basic types of nuclear reactor used to generate power. The first (and by far the most commonly used) is the pressurised water-cooled reactor (PWR) in which ^{238}U enriched with ^{235}U (commonly called enriched uranium; natural uranium is mainly ^{238}U, containing only 0.7 per cent of ^{235}U) in the form of its oxide is 'burnt'. The second, the advanced gas-cooled reactor, is similar but is gas cooled. The third, MAGNOX, utilises non-enriched uranium metal rods as fuel. The final reactor type is the fast-breeder which uses ^{239}Pu as a fuel surrounded by a blanket of ^{238}U which generates ^{239}Pu. The fast-breeder reactor is controversial because it generates more plutonium than it uses up as fuel.

If fast-breeder technology is not pursued the plutonium produced in the other reactor types could be regarded as a waste product of the power generation industry; however, it is possible to generate power from ^{239}Pu in combination with ^{235}U (in so-called mixed oxide fuel (MOX)) in PWR reactors. It is likely that in years to come plutonium will become an important and, in the event that fast-breeder technology is pursued, relatively cheap fuel for the generation of nuclear power. The problem is that the plutonium generated by fast-breeder reactors might also be used in the weapons industry.

7.3.1   *Environmental Impact of Nuclear Power Stations*

There are two ways in which nuclear power stations can adversely affect the environment. Their greatest impact is commonly thought to be the emission of radioactive waste into the air, waterways and by fallout onto land. It is much more likely, however, that the greater impact is from reactors that are built on river estuaries or the sea in order to utilise large volumes of water in their cooling systems. The water pouring out of the power stations is warm and therefore warms the immediate environs of the power station. This warming results in changes in ecosystems. Some species enjoy the raised temperatures and flourish, others die because they cannot tolerate raised environmental temperatures. Alien species might even move to the warmer environment and displace established members of the local ecosystems. Whichever of these effects occurs, a major change to the local environment results from the presence of a water-cooled nuclear power station. Any such change is undesirable because it inevitably means a reduction in the population of the less tolerant species.

7.3.2 Emissions

Despite the relative unimportance of radioactive emissions from nuclear power stations, *vis-à-vis* warming of local environments, emissions remain a very controversial issue. In most countries where nuclear power is generated there are monitoring schemes to check radioactive contamination of the environment. One such scheme operated by the UK Government is the Terrestrial Radioactivity Monitoring Programme (TRAMP) which was set up in 1986 immediately after the Chernobyl disaster as a means by which fallout from Chernobyl could be monitored in the UK with a view to protecting the consumer.

It is very difficult to take sufficient and representative samples of soil and herbage from across the UK to allow a meaningful study to determine the dispersion of and contamination of land with radioisotopes originating from the activity of the nuclear industry. Very cleverly TRAMP samples milk from cattle grazing relatively large areas of land and uses the cattle as a mechanism of sampling and concentrating radioactivity deposited on grassland. The cow acts as a sort of affinity chromatography column, the radioactivity being eluted in her milk. A large number of milk samples are taken from sites throughout the UK and a broad array of radioisotopes measured in concentrated and ashed samples of the milk (this process is necessary because the amounts of radioactivity concerned are so small that very significant concentration is necessary to allow their detection; even after these very significant concentration factors sophisticated background corrections are necessary). This monitoring programme allows the human food chain to be monitored, but also monitors the activity of the UK's nuclear power industry. It is reassuring that during the period of operation of TRAMP there have been no environmentally threatening leakages of radioisotopes from the generating plants.

7.4 Fate of Radioisotopes in the Environment

When radioactivity is released into the environment its fate and behaviour depend upon the chemistry of the isotope. Most isotopes used commercially are metals and are positively charged. It is therefore possible to generalise about their fate and behaviour in the environment. Other isotopes are used in medical diagnosis and treatment and in research laboratories (e.g. ^{131}I, ^{14}C, ^{35}S, ^{32}P). They are generally disposed of via the sewers; however, the quantities concerned are tiny and therefore they have minimal environmental impact. Radioactive isotopes originating from the nuclear industry (including weapons manufacture and testing) constitute by far the greatest proportion of the earth's nuclear pollution load and therefore only these metallic isotopes will be considered here.

If a positively charged isotope is released into an estuary ecosystem (this is the most likely eventuality because of the need to build nuclear power stations near large sources of water) the isotope will bind to negatively charged silt particles (e.g. clay). This binding process sequesters the isotope in the locality of the power station. This is a potential problem because it results in the build-up of radioactive environmental contamination in the water courses surrounding the power station. If the isotope did not bind it would disperse and be diluted infinitely by the sea and so (in theory) have a lower environmental impact. Having said this, there is no evidence of any environmental impact of sequestered radioisotopes in the vicinity of nuclear power stations. We must, however, be very careful not to confuse no evidence with no effect. It is imperative that emissions are kept to a minimum because it is likely that in years to come nuclear power will assume greater importance in the generation of electricity and therefore emissions will increase, albeit not proportionately.

Some animals and plants concentrate specific isotopes, probably because they bear a physicochemical similarity to an element required by the organism. A good example of this is the concentration of ^{137}Cs by lichens and fungi. It is possible that fungi and lichens 'mistake' caesium for potassium. ^{137}Cs was released by the Chernobyl disaster and lichens and fungi in the northernmost reaches of Europe (Norway, Finland and Sweden) concentrated the ^{137}Cs to such an extent that they were dangerous to eat. The reindeer, who eat very specific lichens, continued to eat the lichen despite its high radioactive content. The reindeer, in turn, accumulated the ^{137}Cs and they too became very heavily contaminated. People were advised not to eat reindeer originating from the most heavily contaminated regions. This is a typical food chain concentration effect. There are likely to be many more examples of which we are not aware. The impact of these effects upon ecosystems is impossible to predict, but genotoxic effects might manifest themselves as teratogenic events and tumourigenesis.

It is important not to assume that emissions of radioisotopes only result from the activities of the nuclear industry. Burning coal releases polonium (^{210}Po) which is a component part of this fuel. The ^{210}Po activity on grassland around coal-fired power stations is often greater than the corresponding levels of radioactivity from around nuclear power stations. This is an example of background radioactivity being concentrated and deposited by human activity.

7.5 Chernobyl

On 26 April 1986 a group of Russian power station workers decided to override normal safety procedures in the control room of one of the reactors of the Chernobyl nuclear power plant. This was certainly the worst decision that they had ever made and arguably the worst decision (from the point of view of the global environment) that anyone has ever made. An explosion (thought to have been caused by hydrogen build-up) in the core of the reactor blew off the huge biological shield covering and left the contents of the reactor open to the environment. The fire that followed the explosion continued to release radioactivity for several days. As a result of this course of events some 2×10^{18} Bq of radioactivity were released from the ailing reactor.

At first the Russians did not admit that an accident had occurred (1986 was well within the Cold War and relations between the USSR and the rest of the world were at a low ebb). Soon, however, the incident was detected by radioactivity monitoring programmes in northern Europe. The magnitude of the problem became clear when the winds from the east transported enormous amounts of radioactivity in rain clouds to Finland, Norway, Sweden, north Germany, northern England and Scotland. It rained and radioactivity levels on the ground suddenly increased many thousands of times above background. There was very grave concern at government level because it was not possible to predict what might happen because the Russians refused to give any information about the event. Fortunately the weather pattern was such that the vast majority of the emissions were deposited in the seas of the northern globe.

Despite the weather patterns helping to avert a disaster the like of which it is not possible to conceive, Chernobyl was a major disaster of global proportions and it resulted in major land contamination and increased levels of radioactivity in the human food chain. It is not possible in an overview chapter such as this to cover in detail all of the contaminations that occurred and therefore the problems of the north of England and northern Scandinavia only will be used to illustrate these issues.

7.5.1 *Contaminated Pasture in the North of England*

The northernmost part of England is composed to a large extent of high moor land and is ostensibly a sheep farming area. It soon became apparent that the sheep grazing areas had become contaminated with ^{137}Cs ($t_{1/2}$ = 30 years; β- and γ-emission), ^{134}Cs ($t_{1/2}$ = 2.3 years; β- and γ-emission), ^{131}I ($t_{1/2}$ = 8 days; β- and γ-emission) and ^{90}Sr ($t_{1/2}$ = 28.5 years; β-emission only) following the Chernobyl disaster. The geology and soil of this upland area was mainly limestone bedrock with a very shallow peaty (i.e. acid) soil. This sealed the fate and behaviour of the metallic isotopes because they were not adsorbed by the soil and therefore were readily available to be taken up by plants. The plants were then eaten by animals (e.g. sheep) and residues of the isotopes were deposited in the tissues of the

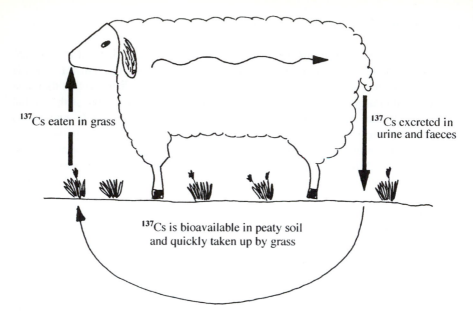

Figure 7.2 Cycling of caesium isotopes in upland peaty soil environments following the Chernobyl disaster.

animals. Immediately after the Chernobyl disaster the tissue levels were sufficiently high for the UK Government to ban their entering the human food chain and it introduced a monitoring programme to ascertain when sheep were safe to be slaughtered for human consumption. The situation would have been very different if the soil had contained clay because the negatively charged clay particles would have adsorbed the positively charged radioactive metals and sequestered them, so preventing plants taking them up and making them accessible to the human food chain. If this had been the case the soil would have remained radioactive for many years; at least the sheep eating the contaminated grass meant that the land contamination decreased because the radioactivity was concentrated in the animals' tissues. The problem was that caesium (particularly) isotopes are excreted in urine so enabling the whole process to start again (see Figure 7.2).

In addition to considering the fate and behaviour of the isotopes which contaminated land immediately after the Chernobyl disaster, it is important to consider the individual isotopes' fate and behaviour in the environment and in animals and their relative 'toxic' risks.

7.5.1.1 ^{137}Cs and ^{134}Cs

The fate and behaviour of both of these isotopes is identical; however, their risks are very different because of their different half-lives. ^{137}Cs's half-life is 30 years compared with a 2.3-year half-life for ^{134}Cs. Both have similar energy emissions, therefore the risk from ^{137}Cs contamination is significantly greater than from ^{134}Cs.

As discussed above, if the caesium isotopes are bioavailable they will be ingested by herbivores as a component part of the plants in their diets.

Caesium is an alkali metal and therefore mimics sodium and potassium in the body. Its salts are very water soluble and for this reason it tends not to reside for long periods of time in tissues, but rather is cleared by the kidney and excreted in urine. For these reasons the caesium isotopes do not present a long-term problem to the consumer. As discussed above, they recycle in the environment and therefore it is possible that the consumer could take in a long-term low dose with their food and, even though the isotopes are not concentrated in the body, the consumer would receive a reasonably high dose of radioactivity via the diet.

7.5.1.2 ^{90}Sr

Strontium is an alkaline earth metal and is a member of the same group of the Periodic Table as calcium and magnesium. In the body it mimics both of these metals being absorbed on the same carrier systems and tending to localise in bone where it is incorporated into the bone's structure and therefore can reside there for many years. This is a significant problem because it results in the bone being irradiated. The long-term effect of ^{90}Sr in humans and other mammals is bone cancer.

Strontium's analogy to calcium means that it is secreted in milk. Cattle grazing land contaminated by Chernobyl fallout produced milk containing ^{90}Sr. This is a particular problem because children drink milk and therefore could lay ^{90}Sr down in their bones. They would be expected to live for longer than adults and so their potential dose from the ^{90}Sr decay is greater and their chance of developing bone cancer is greater.

7.5.1.3 ^{131}I

Iodine is an important element because it is a component of the thyroid hormone thyroxine (see Figure 7.3). If ^{131}I is absorbed into a mammal it is localised in the thyroid very quickly. This is a problem because it means that there is a highly concentrated area of radioactivity that irradiates the surrounding tissue. High levels of radioactivity in the thyroid result in two major effects. In the short term (i.e. acute case) thyroid tissue death occurs and the thyroid output decreases; reduced thyroxine results in reduced metabolic activity which in turn causes lethargy. In medicine parenteral ^{131}I administration is used to treat hyperthyroidism (i.e. overactive thyroid) for just these reasons. In the long term (i.e. chronic case) thyroid cancer might develop.

^{131}I was not too great a problem after the Chernobyl disaster because of its short half-life (8 days). Most of it had decayed before the fallout reached northern Europe. In the immediate environs of Chernobyl itself it is likely to have had a more severe effect; indeed, the incidence of thyroid cancer and birth defects has increased in people from towns near to the power station.

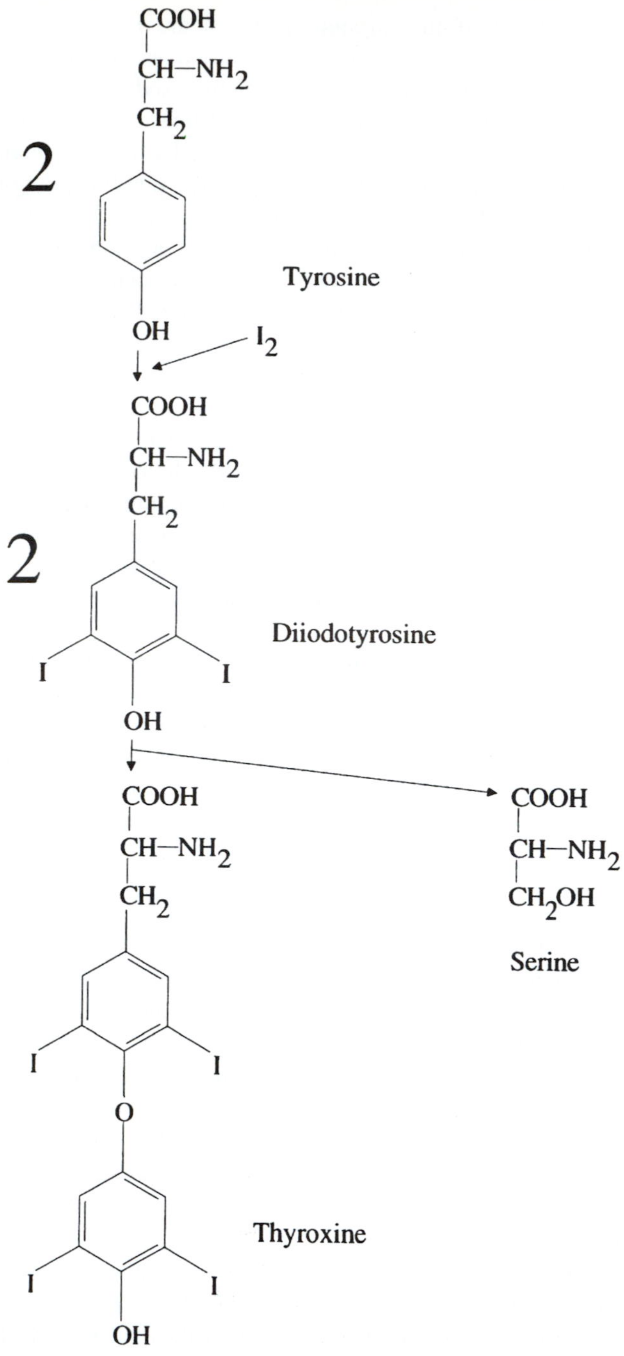

Figure 7.3 Biosynthesis of thyroxine showing the incorporation of iodine.

7.6 Toxic Effects of Radioactivity

7.6.1 α-Emitting Isotopes

The toxicity of radioactive emissions is dependent upon their energy and penetrative powers. The worst cellular eventuality is to have an α-emitting isotope within the nucleus. If this happens the DNA can be bombarded directly by the very high energy, but low penetrative power, α-particle. This is akin to throwing a cricket ball at high speed at a gently balanced deck of playing cards. The DNA damage might result in loss of whole sections of the DNA molecule or fragmentation of the chromosome which would result in serious misreading of the information encoded by the DNA. The cell will attempt to repair this damage, which might be successful, but it could introduce further mistakes and accentuate the problem. The likely eventuality is serious malfunction of the cell which might lead to removal of the mechanisms which control the cell cycle. This is the basis of carcinogenesis (i.e. chronic toxicity). The damage might be so severe that the cell simply dies (i.e. acute toxicity).

If the α-emitting isotope happens to be in a developing embryo the cell damage caused might result in teratogenesis or embryotoxicity (see Chapter 5). The point in the development of the embryo or foetus at which exposure occurs will determine which organs or parts of the body are affected.

α-Emitting isotopes must be absorbed into the body to manifest toxicity. Their penetrative powers are far too weak for them to have major deleterious effects if they are deposited on the surface of the body or if a person, animal or plant is exposed to them in the form of a remote source.

It is important to remember that not only people succumb to the effects of radioisotopes; plants and animals will be affected also. We know far less about these effects in plants than in animals, but mutations will almost certainly result which might affect the viability of the offspring and so have major effects upon ecosystems in the long term.

7.6.2 β- and γ-Emitting Isotopes

The energies of both β-particles and γ-rays are considerably lower than α-particles, but they are very much more penetrative and so can affect cells even from a remote source. For example, if you sat several metres away from a ^{60}Co source the γ-rays would reach you and pass straight through your body.

When either β-particles or γ-rays pass through a cell they cause reactions (often with water) which result in the generation of free radicals (see Figure 7.4). Free radicals are very reactive species which in turn interact with biological molecules causing damage to the molecules and adversely affecting the cell. For example, if the free radicals were generated in the nucleus, damage to DNA might result. The effects would then be similar to those described for the α-emitting nuclides; however, the damage is less severe. A typical result of oxygen free radical attack of

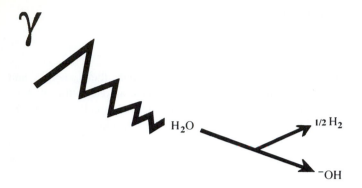

Figure 7.4 Generation of the highly reactive hydroxyl free radical by the passage of γ-rays through water in a cell.

DNA is the 8-hydroxylation of the DNA base guanine (see Figure 7.5). This modification results in guanine being misread as thymine and so proteins synthesised from free radical damaged DNA have the wrong amino acids within their structures. This, of course, can significantly affect their activities (e.g. if the amino acid concerned is within the active site of an enzyme, loss of enzyme activity might result).

Another possible mechanism of damage to biological molecules by radioactivity is restricted to β-emitting isotopes. β-Particles are electrons and therefore carry a negative charge. If a negative region of a molecule is hit by a β-particle it might knock off a fragment of the molecule because of the charge repulsion between the negative region of the molecule and the electron (this is similar to fragmentation in the mass spectrometer (see Chapter 4)). This is akin to a tiny stone being catapulted at a finely balanced deck of cards and the top card being shot away. Molecular groups such as the carboxyl moiety (-COOH) are candidates for removal by β-particle bombardment.

7.6.3 *Leukaemia Clusters*

As discussed above bombarding cells (and particularly the nucleus) with radioactivity leads to changes which might result in removal of the normal control mechanisms of the cell, which in turn might lead to the development of cancer. It has long been suspected that the incidence of cancer *should* be greater in the immediate environs of nuclear power stations because presumably people living near to such stations are receiving greater doses of radioactivity than the general population. This increased dose could be for one of two reasons: that the environment is contaminated and therefore simply by living in the area their exposure to radioisotopes is greater or that a great proportion of the inhabitants of the locality of nuclear power stations work at the power station (this is very likely

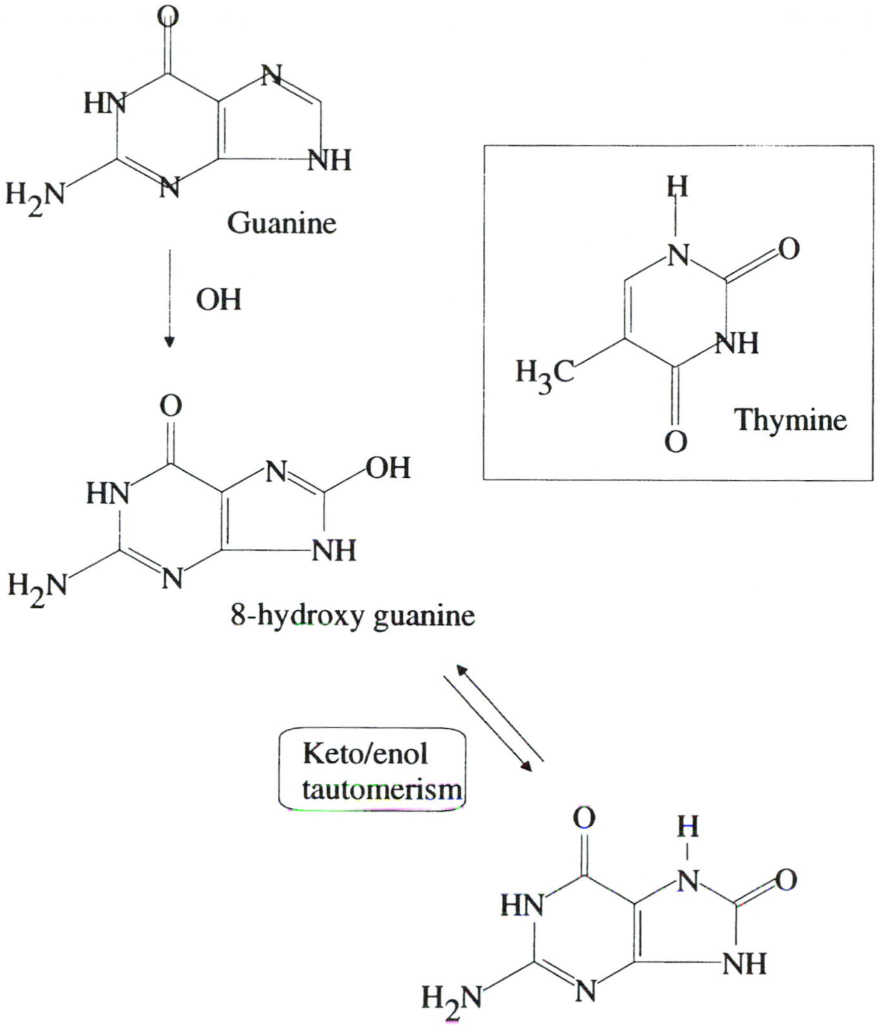

Figure 7.5 Free radical hydroxylation of the DNA base guanine to form 8-hydroxyguanine which is misread by the DNA replicative apparatus as thymine (inset).

because the power stations are usually in remote areas and people move to these areas specifically for work) and therefore are exposed in the workplace.

During the mid-1990s epidemiological studies showed that there was a marginally increased incidence of a specific very rare leukaemia in the families of workers at the Sellafield site (this site generates nuclear power and reprocesses fuel) in Cumbria in the northwest of England. Exposure to radioactivity at work, or workers bringing radioactivity home on their clothing, was blamed at first and the press made a great issue of this. Since the initial release of the report the furore has died down and other possible causes of the leukaemia around Sellafield have been proposed. For example, it is possible that the leukaemia is caused by a virus and that the virus is present in the relatively remote Sellafield community. It is by no means proved that this particular cluster was caused by workers' exposure to radioactivity at work or via environmental contamination.

7.6.4 *Radon*

Radon is a naturally occurring (4×10^{-17} per cent by weight of the earth's crust) radioactive gas containing an array of isotopes (^{220}Rn, $t_{1/2}$ = 55 seconds; ^{222}Rn, $t_{1/2}$ = 3.8 days (this is the environmentally important isotope); ^{219}Rn, $t_{1/2}$ = 4 seconds). They all decay by α-emission.

Radon is formed by the decay of the ^{238}U and ^{235}U series (see Table 7.6). It has risen in prominence in the late 1990s because of its occurrence in the atmosphere of public and domestic buildings, especially in Scandinavia. The reason for this sudden upsurge in interest in radon is because of our environmental awareness and our desire to keep our heating bills low which has led to the installation of efficient insulation systems (e.g. double glazing). This has resulted in a very significantly reduced air exchange in many homes. If houses are built on granite bed rock or have used granite in their construction and are well insulated a build-up in radon might occur; the occupants are therefore exposed to higher doses of radon than people in draughty houses. The Scandinavians, because of their very cold winter climate, have particularly well-insulated homes, and it is for this reason that the radon problem appears to be greater in these countries.

The worst outcome of chronic exposure to radon is lung cancer. This could result from continuous respiratory exposure to the radon gas in the atmosphere or because the radon adsorbs onto minute dust particles ($>5 \mu$m in diameter) which are inspired and remain deep in the lung where they irradiate the surrounding tissue.

$$
\begin{array}{lll}
^{220}\text{Rn} \rightarrow & \alpha & 55.3\text{s} \\
\Downarrow & & \\
^{216}\text{Po} \rightarrow & \alpha & 3.04 \times 10^{-7}\text{s} \\
\Downarrow & & \\
^{212}\text{Pb} \rightarrow & \beta & 10.64\text{h} \\
\Downarrow & & \\
^{212}\text{Bi} \rightarrow & \alpha & 60.6 \text{ min} \\
\Downarrow & & \\
^{208}\text{Tl} & & \\
\Downarrow & & \\
\end{array}
$$

$\beta \leftarrow$ $\Downarrow$

25s $\alpha \leftarrow ^{212}$Po

$\Downarrow$

^{208}Pb (stable) $\Leftarrow \quad \Leftarrow \quad \Leftarrow \quad \rightarrow \quad \alpha \quad 3.1$ min

Figure 7.6 Radioactive decay chains for the radon isotope ^{220}Rn.

7.7 War

The most controversial use of radioisotopes is in nuclear weapons. The association of radioactivity with nuclear weapons is one of the prime reasons that many people are against nuclear power. Of course their worries are, to some extent, well founded because the plutonium produced by fast-breeder nuclear reactors might well find its way into the nuclear weapons industry.

It is far outside the scope of this book to consider fully the environmental impact of the nuclear weapons industry and the use of nuclear weapons in war, but for completeness a very brief review of the subject will be undertaken.

We can have no real idea of the impact of modern nuclear weapons upon the environment and human populations. Our only benchmark is the unbelievable carnage and devastation caused by the bombs dropped on Hiroshima and Nagasaki in 1945. Since then the power and size of nuclear weapons has increased astronomically.

The effects of nuclear weapons are fourfold. Their manufacture and testing have an environmental impact. Manufacture results in emissions (albeit small; see Table 7.5) into the environment. Testing releases large amounts of radioactivity into localised environments and destroys geological and ecological structure. In 1996 there was very great controversy about the French nuclear weapons testing programme at Moraroya Atoll in the Pacific Ocean. The explosions set off deep in

Table 7.5 Radioactivity in lettuce grown near the Aldermaston nuclear weapons site in southeast England, showing the extremely low levels present

Isotope	Level (Bq kg^{-1})
^{3}H	< 20
^{60}Co	< 0.7
^{95}Zr	< 1.9
^{95}Nb	< 2.0
^{103}Ru	< 1.6
^{110}Ag	< 0.9
^{134}Cs	< 0.5
^{137}Cs	< 0.7
^{144}Ce	< 3.0
^{238}Pu	$< 2 \times 10^{-4}$
^{239}Pu + ^{240}Pu	7×10^{-4}
^{241}Am	7×10^{-4}
Total uranium	0.2

This shows that confinement within the nuclear site is very effective.
Data from the TRAMP report for 1992, Ministry of Agriculture, Fisheries and Food, Published by HMSO, London.

the bedrock beneath the sea resulted in shock waves of such force that coral reefs were completely destroyed and presumably animals in the area were killed. The direct impact of exploding a bomb during war is obvious – unbelievable devastation and radioactive contamination of the surrounding area and later transport and fallout of radioactivity in geographically remote areas. The final impact is far less obvious. When a nuclear bomb is exploded it causes an enormous dust cloud to rise high into the atmosphere, this spreads with the prevailing weather system. The dust particles reflect the sun's light and heat and this results in cooling of the earth.

Further Reading

Malcolme-Lawes, D.J., 1979, *Introduction to Radiochemistry*, London: Macmillan.

Savchenko, V.K., 1995, *The Ecology of the Chernobyl Disaster*, Carnforth, UK: Parthenon Publishing Group.

Woodhouse, J. and Shaw, I.C., 1996, Plutonium — an element of our times, *Toxicology and Ecotoxicology News*, **3**, 36–42.

8

Legislation

Legislation is crucial in controlling pollution. The process of legislation and the environmental Directives and laws are discussed in this chapter:

- Dangerous Substances Directive
- Plant Protection Products Directive
- Biocides Directive
- Food and Environment Act
- Control of Pesticides Regulations
- Wildlife and Countryside Act
- Water Resources Act

A brief overview of legislation outside the European Union is included.

8.1 Overview of Environmental Legislation

For hundreds of years we had accepted the growth of pollution and only when it became utterly intolerable did we take effective action against it. Now we are seeing how much it has deprived us ... year by year we shall need to revise our policies so that our economy can grow in a way which does not cheat on our children.

John Gummer
Secretary of State for the Environment in the UK
(extracts from *An Introduction to Sustainable Development, The UK Strategy 1994*) 'After Rio'

As human populations have grown over the millennia we have become increasingly aware of our environment and our capacity to threaten the very future of humanity. Without a collective and unified effort on the part of all nations we have the real capacity to destroy a significant part of our environment and to threaten civilisation itself. In the past, through ignorance and over exploitation, we have seen the extinction of many species, notably the dodo and the American carrier pigeon in the nineteenth century and very nearly the North American bison. The list is endless. Stone age man cleared forests and drove the wolf, the bear and many other species to extinction in parts of Europe. In the past, early hunters could not have been expected to know the consequences of their actions. It was perhaps less forgivable for early settlers to massacre the carrier pigeon in millions to be transported by the trainload to be eaten as a delicacy on the tables of the rich. Like the bison, the pigeons congregated in such large numbers that hunters assumed there was an endless supply. In reality these are communal animals that lived in only a few very large flocks and herds and were very vulnerable to large-scale slaughter. Only the bison survived. The fact that extinctions occur is not really important. Many species that existed in the past, like the dinosaur, have died out from natural causes. Many species will die out from natural causes in the future. There is no way that we can accurately record all the species that ever existed or even the number of species that exist today. In general, however, biological diversity has increased slowly over geological time with occasional setbacks through mass global extinctions.

We can only speculate on these previous extinctions which are likely to have been caused by catastrophic events such as giant meteors striking the earth. The last major extinction was tens of millions of years ago; however, what is becoming more certain is that a further major decline is now underway as a result of human activity. A significant element of the decline is likely to be due to the introduction of chemicals into the environment. Until legislation was introduced in the 1960s and 1970s our birds of prey were threatened by the use of the pesticide DDT. More recently in the European Union (EU), bat populations have been in decline. This has been attributed to the use of persistent organochlorines in wood preservative products. Other chemicals such as TBTO have caused molluscs to permanently change sex and a bank of other oestrogenic chemicals has been implicated in sex changes in fish and a 50 per cent reduction in the sperm count of western men. Without legislation, and more importantly the power to enforce it, the release of chemicals into the environment can go unchecked with potentially disastrous consequences.

Many chemicals, especially pesticides, are recognised as having high biological activity. If they are to be used and disposed of safely, legislation and controls must be imposed on the manufacturers and users of such chemicals. These controls must embody a philosophy of 'cradle to grave' custody of a compound. For instance, we should not only consider the effects of copper chrome arsenic in treated timber escaping from a treatment plant, we must also consider the end use of the timber in children's playgrounds and on motorway fencing. We must also consider what happens to the wood and the chemicals in it at the end of its effective life when the timber is burned in the open or on domestic fires or in power stations. We must consider what happens to those residues of chemicals which pass via smoke into the atmosphere. Depending on the persistence of a chemical and its toxicity, this control may involve containment for thousands of years. In the case of radioactive substances, whose half-life can be measured in hundreds or thousands of years, we have to predict the movement and stability of these chemicals over this time period and also be assured that our technology is capable of containing this highly toxic waste intact over this period.

Legislation must ensure that controls bite at an international level. We have seen many countries adopt a 'not in our back yard' philosophy to waste disposal but think little of exporting their pollution for disposal in the Third World. Too often in the past our rivers have been open sewers for the disposal of industrial waste and pesticide residues. Within Europe this has led to the North Sea becoming one of the most polluted seas in the world. Thankfully, all this is changing and nations are increasingly uniting as more responsible custodians of our environment.

Current environmental legislation is open to interpretation, not only at a national level but at several levels internationally where laws, language and customs differ. Compliance with internationally made laws is often open to interpretation. Furthermore, in order to bring about new legislation a complicated series of official steps needs to be taken both nationally and internationally. This can take many years and the international scene is littered with failed and abortive pieces of legislation. The final legislation, even if successful, often bears little resemblance to its original form as a result of the collective bargaining activity employed by different nations to obtain political self-interest in key areas. Many good, hard hitting pieces of legislation have, however, emerged from this process within recent years. Perhaps the most significant was The Single European Act 1986 which had the effect of putting environmental issues high on the list of priority for reform. Many subsequent Directives have had a strong environmental element and Europe can be seen as leading the world in environmental reform.

8.2   European Legislation

The EU has developed a general chemical policy embodied in a range of different Directives which, compared with previous chemical control, are innovative and should bring about real improvements to our environment. The EU has also developed a policy for the control of pollution in all its member states. Currently it is seeking to link chemical with pollution policy and expand the remit of its Directives to encompass risk as a concept, in addition to hazard. It is hoped that we will eventually have an integrated substance-based approach to chemical control which will control pollution from whatever source. The Americans have also adopted a general chemical and pollution policy coupled with legislation which in many ways parallels the European model. National schemes also exist in many countries, often side by side with international controls. To understand the complexity of legislation throughout the world, however, it is perhaps best to explain the processes of law and legislation at a national and international level.

To understand the European dimension of EU law and its institutions we must first discuss its structure. There are four major EU institutions which allow the process of legislation under the Treaty of Rome. The Treaty of Rome was set up in 1957 and defines the main aims of the sovereign member states to establish a common market which expands economic activity, promotes the standard of living and encourages stability. Its main purpose was to create a single 'common' market without internal barriers between member states. The treaty provides the legislative framework by which this can be brought about. The four major institutions are:

The Commission
The European Parliament
The Council of Ministers
The European Court of Justice.

8.2.1   The Commission

The Commission is the supreme EU executive comprising of 20 independent members appointed by individual member states: two each from Germany, France, the UK, Spain, and Italy and one each from the remaining member states of Belgium, Denmark, Greece, Ireland, Luxembourg, The Netherlands, Portugal, Sweden, Austria and Finland. These members act independently in the interests of the community as a whole, and not as national representatives. Each commissioner has responsibility for an area of community policy. Their main function is to propose EU legislation. Operationally, it is divided into 25 different 'Directorates' which meet in secret. These report to the Council of Ministers. The most important Directorates concerning chemical and environmental legislation are:

Directorate General vi Agriculture
Directorate General xi Environment, Nuclear Safety and Civil Protection
Directorate General iii Industry.

8.2.2   The European Parliament

The European Parliament is a consultative and advisory body. As such it is the only open part of the EU decision making process. Its function is to 'assess proposals for legislation'. It considers and comments upon the Commission's proposals and has some control over community budgets. The 626 Members of the European Parliament (MEPs) are elected for a 5-year term by citizens of the 15 member states. The UK, France, Germany and Italy have the most MEPs whereas smaller member states have allocations according to the population size of the country (see Table 8.1). If the Council of Ministers does not consult with the European Parliament all legislation is 'void'.

8.2.3   The Council of Ministers

The Council of Ministers is composed of government ministers from each member state and as such is a political body and the first decision making organ of the EU. Its main function is to consider proposals from the Commission and each minister in turn holds the presidency of the Council for a period of 6 months, during which

Table 8.1 The number of MEPs elected by each member state

Germany	99	Greece	25
France	87	Portugal	25
Italy	87	Sweden	22
United Kingdom	87	Austria	21
Spain	64	Denmark	16
The Netherlands	31	Finland	16
Belguim	25	Ireland	15
Luxembourg	6		

time they chair all meetings. The President considers proposals from the Commission or Council/Presidency initiatives. All decisions must be unanimous for proposals to be accepted.

8.2.4 *The European Court of Justice*

The European Court of Justice is based in Luxembourg. It consists of 15 judges appointed by the member states. The main function of the Court is to interpret and apply all community law, the basic treaties under the Treaty of Rome to specific regulations, Directives and decisions issued by the Council and the Commission. All judgements are binding and member states must ensure that any legislation is compatible with its own national laws. In the UK the rights and obligations granted or imposed by the treaties of the European Union are incorporated into UK law by the European Communities Act 1972 and are therefore binding for most legislation. Under the Treaty of Rome Section 2, however, EU law is not enforceable in the UK for environmental issues without special legislation to implement them. Thus in the UK, environmental legislation brought about in Europe is deferred. The process of legislation is summarised in Figure 8.1.

Community law is contained in treaties under the Treaty of Rome 1957 or the Single European Act 1986. Ninety per cent of community law made by the Commission is made on common agricultural policy. The main legislation to emerge from the Commission is in the form of Directives. These Directives are normally enforceable across all member states and are the most important source of UK legislation. They have been responsible for most of the statutory controls in environmental legislation.

There are two main types of Directive. The 'key' Directive is called a 'Framework Directive'. These Directives impose a raft of controls which can be annexed into 'daughter Directives'. Member states are required to take all appropriate measures to ensure fulfilment of obligations arising from this treaty. Article 100 enables Directives to be issued for approximation or harmonisation of laws across member states. The Directives are given a number and it is common to refer to the number rather than the title of the Directive.

THE PROCESS OF LEGISLATION

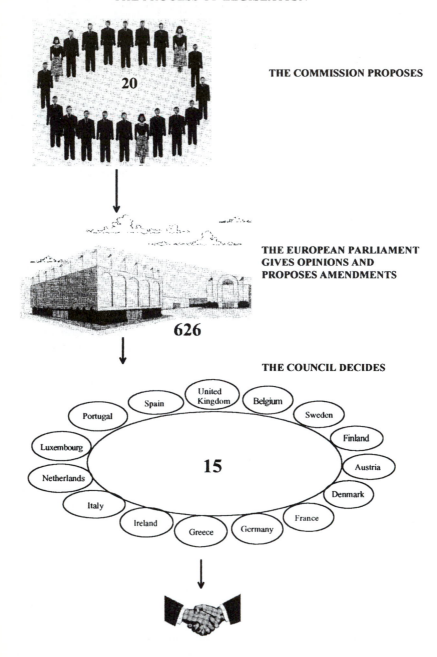

THE COMMISSION PROPOSES

THE EUROPEAN PARLIAMENT
GIVES OPINIONS AND
PROPOSES AMENDMENTS

THE COUNCIL DECIDES

Figure 8.1 Schematic representation of the process of legislation.

The majority of existing legislation in the EU on chemicals relates to controls on active ingredients or 'substances', although moves are now being made to bring the environmental assessment of preparations under EU law. The main legislation on chemicals is discussed below.

8.3 New Chemicals

The Dangerous Substances Directive 67/548 EEC
New chemicals were brought under European legislation under the Dangerous Substances Directive 67/548 in 1967. (The prefix to the second number normally denotes the date of adoption.) The purpose of the first Directive was to make sure that new chemicals that came onto the market in sufficient volumes to be considered a hazard to humans or the environment were given a classification and a hazard label when being transported into or within countries in order that manufacturers, transporters and the public were aware of the dangers of the chemical if released. The legislation at the time only really covered high volume of supply chemicals. These regulations have been amended a number of times in order to improve the hazard assessment. An assessment of the substance is made based on data relating to degradation, biodegradation, hydrolysis, photo-degradation, persistence and toxicity to organisms as representatives of different trophic levels. The data requirements under the scheme are dependent on tonnage used. For instance, under a limited announcement where there is a limited use of between 0.1 and 1 ton per annum, ready biodegradability and maybe information on the toxicity to sewage systems (microorganisms) may be required. Acute toxicity to *Daphnia* may now also be required, as under the more recent 7th amendment to 92/32/EEC a risk assessment as well as a hazard assessment is required.

8.3.1 *The 7th Amendment to 67/548*

The recent (1992) 7th amendment 92/32/EEC to the Dangerous Substances Directive now demands testing on different trophic levels, more fate and behaviour data and requires the competent authority (the regulation authority in different member states) or the notifier to carry out a risk assessment to assess the likelihood of the chemical reaching the environment. Under this scheme, if the chemical is supplied in amounts greater than 1 ton per annum then other tests become mandatory. For instance, ready biodegradability/abiotic degradation, acute toxicity to *Daphnia* and fish, algal growth inhibition tests and soil adsorption/desorption screening tests are required over and above details on physicochemical properties. Soil adsorption and algae tests are recent additions to the data requirements under the 7th amendment. These are known as base level tests. The 7th amendment also seeks to further harmonise classification in member states. Under the 7th amendment the assessment of low volume chemicals was harmonised at EU level for the first time. If there is a probability of the chemical being supplied at up to 500 tons per annum in total then the notifier must progress to a higher level of testing. This brings in the added requirement of inherent biodegradability tests and prolonged toxicity to *Daphnia* and fish. If tonnage is likely to be up to 5000 tons per annum in total then additional biodegradability and mobility tests along with prolonged acute and subacute toxicity to birds are required. These are known as stage II tests. The regulators in each member state make an assessment based on this data leading ultimately to the classification and labelling of the substance under Council Directive 93/21/EEC. EU legislation on the classification, packaging and labelling of dangerous substances requires that chemicals marketed in the EU must be classified according to their hazardous properties. An additional classification, 'dangerous to the environment', based largely on aquatic toxicity, has recently been introduced. Chemicals for supply are labelled with an 'N' symbol which denotes if they are toxic, very toxic or harmful to the environment (see Figure 8.2).

DANGEROUS TO THE ENVIRONMENT

Figure 8.2   EU 'N' symbol used to label chemicals that are toxic to the environment.

8.3.2 Classification and Labelling

The system for classification and labelling of dangerous substances in the EU is based on Council Directive 67/548/EEC which, as we have explained, has recently been updated by the 7th amendment. The Directive and all its subsequent amendments lay down a system for classification and labelling based on criteria covering a wide range of potential dangerous effects (see Section 8.3.3). The category 'dangerous to the environment' is used for substances and preparations which present or may present immediate or delayed risk for one or several compartments of the environment. Once a substance has been classified in one or more categories a label follows automatically.

8.3.3 Hazard Symbols and Phrases

The most severe hazards are highlighted in symbol form. The dangerous properties are denoted by risk phrases (R-phrases) and safety phrases (S-phrases). In the 7th amendment to the Directive the category 'dangerous to the environment' has a new 'N' symbol (see Figure 8.2).

The label, which must be placed on all substances for sale or supply, must draw the attention of persons handling or using it to the dangers involved (you might have seen these symbols on the rear of tankers carrying dangerous loads on the road). All substances must be labelled when marketed. For labelling of substances which are dangerous to the environment, substances are divided into two groups according to their acute and/or long-term effects on non-aquatic systems. The criteria for the second group are still ill-defined; however, the risk phrases which apply for the aquatic environment are better defined and their application is simplified below.

Substances 'dangerous to the environment'

Risk Phrase	Toxicity	Readily Degradable	Symbol 'N'
R50	$EC_{50} < 1 \, mg \, l^{-1}$	Yes	Required
R50, R53	$EC_{50} < 1 \, mg \, l^{-1}$	No	Required
R51, R53	$1 < EC_{50} < 10 \, mg \, l^{-1}$	No	Required
R52, R53	$10 < EC_{50} < 100 \, mg \, l^{-1}$	No	Not required
R52, R53	Undetermined	No	Not required

R50	Very toxic to aquatic organisms
R50, R53	Very toxic to aquatic organisms and may cause long-term adverse effects in the environment.
R51, R53	Toxic to aquatic organisms and may cause long-term adverse effects in the aquatic environment.
R52, R53	Harmful to aquatic organisms and may cause long-term adverse effects.

R52 and R53 are used separately or combined where there is concern about persistence and prolonged toxicity. Substances which bear the risk phrases R52 and/or R53 do not need to be additionally labelled with the symbol 'N', although they are classified as dangerous to the environment and are assigned to appropriate risk phrases.

We have mentioned that criteria for the non-aquatic environment are less well defined. An assessment is currently being made by the competent authority on the basis of toxicity, persistence, potential to bioaccumulate and predicted or observed environmental fate and behaviour of the substance. When these properties present a danger to the function of the natural structure of an ecosystem one of the following risk phrases is assigned:

R54	Toxic to flora
R55	Toxic to fauna
R56	Toxic to soil organisms
R57	Toxic to bees
R58	May cause long-term adverse effects in the environment
R59	Dangerous to the ozone layer.

8.3.4 Risk Assessment

At the present time (1996) no substance has been classified or labelled with regard to its hazard to the environment; however, there is now an additional requirement for substances classified as 'dangerous to the environment' for a risk assessment. The implementation date for this requirement was 1 November 1993. The process is somewhat involved and as this chapter deals predominantly with aspects of legislation and the process of risk assessment is dealt with elsewhere, we shall merely outline the legal requirements which may dictate further testing or revocation of substances.

The risk assessment involves predicting the amount of a substance that will enter the environment and comparing this with acute and prolonged toxicity studies. Predicted environmental concentration (PEC) should be:

1000 × lower than the acute LC_{50}/EC_{50} for the most sensitive species (algae, *Daphnia* or fish)

50 × lower than the lowest NoEC value from prolonged toxicity tests involving at least two taxonomic groups

10 × lower than the lowest NoEC value where data from three taxonomic groups are available.

The concentration for environmental concern (ECC) is thus:

$$\frac{LC_{50}/EC_{50}}{\text{Assessment factor (1000, 50 or 10)}}$$

A substance is of no concern if the PEC is less than the ECC; however, if the value is greater, test requirements may be elevated to the next tonnage level. Specific recommendations may be enforced to reduce this risk, or a notification may be refused and an existing substance withdrawn.

8.3.5 Strengths and Weaknesses

Many sections of industry and even officials within government are not happy with the notification scheme and the imposition of labelling requirements. It places a great burden on the administrative framework of the Commission and of member states. Even though only five Commission staff coordinate the work of the EU notification system, up to 1000 people work on the scheme in the individual member states. Some sections of industry find it particularly burdensome and claim it prevents innovation as it takes vast resources and years to get a substance notified. Many see the scheme as guarding the health of the environment and would be reluctant to see it fundamentally changed. In these days of government deregulation it is perhaps fortunate that deregulation of the scheme is not at the moment on the cards. The scheme must remain transparent and be seen to be working by all those involved. It is now recognised that the new substances scheme is well established and what is currently required is more concentration on the regulation of the many thousands of existing substances (chemicals).

8.4 Existing Substances

There are many thousands of existing substances that have never been subject to notification under the Dangerous Substances Directive; however, the EU requires that basic information is provided on all chemicals produced in, or imported into, the EU in quantities over 10 tons per year. This information, which includes data on the environmental risks, will be used to prioritise these chemicals and to carry out a risk assessment as for new substances. Then necessary control measures will be proposed. The EU existing substances regulations are being fully coordinated with the Organisation for Economic Cooperation and Development (OECD) programme which makes recommendations and draws up guidelines for environmental controls on chemical substances. The OECD is the main coordinating body for the development of new ecotoxicity testing strategies. It agrees appropriate tests which normally end up as mandatory data requirements in EU Directives. In turn, the OECD programme draws on information and expertise from all OECD countries including the USA and Japan.

Although the existing substances scheme bears many similarities to the new substances scheme there are differences. For example, for existing substances it is often possible to draw on monitoring data or discharge data following accidental spillage when conducting risk assessments. In essence, within this scheme there is a greater potential to predict the impact of a chemical on the environment and hence bring about mandatory controls than in the notification of new substances (NONS) scheme.

What is similar to the new substances scheme is the requirement for a risk assessment and for importers to fulfil the base set tests under Annex VII of Directive 67/548/EEC. Under the regulations each member state is required to act as a rapporteur for a number of priority substances which have been recommended for risk assessment. The rapporteur evaluates the information and prepares the risk evaluation. The conclusions drawn from this will involve a process identical to that described for the new substances scheme. The information available on existing substances will differ greatly from that on new substances. Few substances will have base set data tests; these may not have been carried out to GLP (good laboratory practice) and many will be substances supplied at high tonnages. Again, like the new substances scheme, the ESR regulations are designed to be transparent to enable manufacturers and importers of existing substances to anticipate the information that may be required by competent authorities performing risk assessments.

8.5 Other EU Controls

8.5.1 *Restrictions on Marketing and Use*

Since being adopted in 1976 the Marketing and Use Directive (76/769) has been amended 11 times to control especially dangerous chemicals. More recently, in 1995, pentachlorophenol was reviewed under the Marketing and Use Directive because there was concern that the public and the environment could be exposed to harmful residues as a result of its use as a wood preservative. The process of identifying chemicals for scrutiny is on a rather *ad hoc* basis. It normally arises from a single member state having a concern and declaring to the other member states that they are going to ban a chemical. The Commission must then respond. A number of countries believe that decisions to ban chemicals under this route is not done under any strict scientific scrutiny as it only looks at the hazards not the risks of a substance and does not explore the cost benefit analysis of a substance. For instance, there are plans under the Marketing and Use Directive to ban brominated fire retardants in furniture coverings because dioxins may be released if substances are not disposed of correctly. Many lives are saved each year, however, through their use as flame retardants. As a result the European Parliament has refused to give a ruling until this issue is clarified and the proposed Directive cannot therefore proceed. Clearly the mechanisms under the Marketing and Use Directive are far from satisfactory and a more harmonised approach is necessary. At the very least legislation through this route needs to take into consideration not only the hazards of chemicals in the environment but their risk of exposure to humans and the environment and the hazards and risks of substitutes that may be used. The issue of comparative assessment is now recognised as an important consideration for risk reduction from the use of chemicals in the environment. Most important perhaps when considering restrictions under the marketing of chemicals is the need to get a balanced view. There is also an overriding need to encourage the production and use of alternative less harmful chemicals which are less persistent in the environment.

8.6 Specific EU Directives Regulating Pesticides

There are many other EU Directives setting standards for air, soil and water quality, particularly for drinking water, bathing water and water supporting fish. For these, member states must ensure that all sources of input into these environments do not result in statutorily agreed environmental quality standards (EQS) being exceeded. Tributyltin oxide, which is used in wood preservatives and antifouling paints, is known to cause gross malformation in shellfish and an EQS of $20\,\text{ng}\,\text{dm}^{-3}$ is set for fresh water and $2\,\text{ng}\,\text{dm}^{-3}$ for marine environments. These standards cover diffuse as well as point sources from individual installations in shipping sources. This approach is useful for chemicals such as pesticides which are deliberately applied to agricultural land or used in the industrial pretreatment of wood in impregnation plants that are often adjacent to rivers and water bodies. Pesticides are a special case. They are designed to be toxic to living creatures. Pesticides are substances, preparations or organisms used or prepared for destroying pests. They are used to destroy organisms which are harmful to plants, wood, plant products and waterways, buildings, manufactured products and animals, and include herbicides, insecticides, fungicides and molluscides, acaricides, neuroticides and sterilants. Plant growth regulators and animal repellents are also included. In Europe plant protection products have EU-wide legislation under the Plant Protection Products Directive. The non-agricultural pesticides Directive (the Biocides Directive) is still being negotiated. These are both important pieces of chemical legislation and require more in-depth discussion.

8.6.1 *The Plant Protection Products Directive (PPPD)*

The PPPD or EU Directive 91/414/EEC came into force on 15 July 1991. The purpose of the Directive is to harmonise arrangements for the marketing of plant protection products within the community. This is its primary role although the environmental safety of pesticides is particularly addressed. Particular reference is given to the data required for the authorisation of products and their evaluation in order to determine the possible risks to wildlife and the environment that may arise from their use. Many of the data requirements are similar to those required for new substances. For instance, toxicity profiles on fish, *Daphnia*, algae, birds, etc., a risk assessment as to the fate and behaviour of chemicals in the environment, a prediction of likely concentrations which will enter the environment and an assessment of the impact on wildlife and humans must all be carried out. The legislation covers both new and existing pesticides. Annex 1 of the Directive will eventually contain a list of active ingredients which have been reviewed EU wide. These substances will then have approval for use throughout the community.

There are about 8000 active substances on the market throughout the community. The Commission has agreed that an initial 90 should be reviewed first by different countries. The number of reviews carried out by each country depends on their resources to carry out the work, with major countries such as the UK, France and Germany getting 12 each. There was a requirement for member states acting as rapporteurs to have received all the details on the substances by the end of 1994. They were then required to report back to the Commission within 12 months. The review procedure then progresses with information being sent to the European Standing Committee on Plant Health. They may request further information but it is expected that a draft Directive establishing annex 1 will be agreed on some time in 1997. Annex VI of the Directive was adopted as daughter Directive 94/43/EC on 27 July 1994. This is called the Uniform Principles Directive. This establishes the data requirements and the methods for assessing the hazards and risks of the use, sale, storage, etc. of plant protection substances but allows member states some flexibility in their approach.

8.6.2 *The Biocides Directive*

The Commission published a proposal for a Directive for the placing of biocidal products on the market in September 1993. In 1995 common principles were adopted

Table 8.2 Other EU legislation relating to chemicals and pollution control

Directive	Title
R2455/92	Export of Dangerous Chemicals
79/117/EEC	Pesticides – Use Restrictions and Amendments
76/895/EEC	Pesticide Residues
85/501/EEC	Major Accident Hazards (plus 1st and 2nd amendments)
	Seveso Directive
80/778/EEC	Drinking Water
80/68/EEC	Ground Water
76/464/EEC	Dangerous Substances in Water
78/659/EEC	Water Standards for Freshwater Fish
79/923	Shellfish Water
84/360/EEC	Emissions from Industrial Plants
75/442/EEC	Wastewater Framework (amended by 91/156/EEC)
78/319/EEC	Hazardous Waste Directive (replaced by 91/689/EEC)
75/439/EEC	Waste Oil (amended by 87/101/EEC)
86/278/EEC	Sewage Sludge

for the Biocides Directive. The common principles are not dissimilar to the uniform principles in the Plant Protection Directive 91/414/EEC. It contains many similar elements to the Plant Protection Directive but will deal with wood preservatives, rodenticides, insecticides (household and public hygiene), antifouling agents, surface and water biocides, disinfectants, fumigants and preservatives for technical and household materials and preservatives for works of art, as opposed to chemicals used in agriculture. It is likely to establish a community list of authorised substances to be used in products and, like other chemical legislation in Europe, will require data requirements and risk assessments to establish this list. This Directive is currently being discussed in Europe by the European Council of Ministers.

There are a number of other Directives which cover chemical legislation and the prevention of pollution in the EU. As they have been fully integrated into the UK system we discuss these under UK legislation. The main additional Directives are listed in Table 8.2.

8.7   Chemical Controls Outside the EU

The above is a brief overview of chemical legislation in Europe. Chemical and pollution regulations outside the EU have a tendency to rely on computer models to predict environmental effects. This is true of the Environmental Protection Agency (EPA) in the US although there are many similar features to the EU approach. Some states have their own additional registration schemes for pesticides. Important environmental legislation in other countries is discussed below.

8.7.1   *Japan*

In Japan there is no specific regulation for biocides and new substances are regulated under the Industrial Chemical Control Law (ICCL). The scheme is based on an inventory for existing substances and notification for new substances. Three different agencies are involved which slows down the approval system considerably. As a result of cultural differences the authorities can be less amenable to scientific arguments than the EU and America.

8.7.2   *USA*

The USA's system is well established and deserves a special mention. The US control of chemical substances largely follows the European model having both a chemical policy for chemical substances, the Toxic Substances Control Act (TSCA), and the Federal Insecticide, Fungicide and Rodenticide Act (FIFRA). It further controls pollution by setting emission, quality and waste handling standards to protect food, the workplace, air and soil from toxic residues. The TSCA came into force in 1976. It was designed to enable powers to be brought into force to obtain data on the effects of chemical substances and mixtures on health and the environment and to provide the authority to enforce and regulate these standards such that they did not present an unreasonable risk to health and the environment while not impeding unduly and creating unnecessary economic barriers to technical innovation. The US Government has recognised that these initial controls achieved limited success and have over the past 20 years updated these regulations. In 1990 Congress adopted the Pollution Prevention Act which states that 'pollution should be prevented or reduced at source wherever reasonable'. Also in this year they amended the Clean Air Act to encourage local state and federal governments to adopt measures for pollution prevention. The EPA constantly updates its legislation and, like Europe, they are moving away from the risk-based assessments which predict how much pollution the environment can stand to one of prevention of pollution at source. For new chemicals the process of registering chemicals is similar in the USA to that in Europe under the notification of the new substances scheme. Indeed, the USA and the EU regularly compare decisions made on certain chemicals where limited toxicological data are available. The EPA legislation has resulted in a very real impact on the release of toxic chemicals to the environment. Between 1979 and 1990 the EPA banned or restricted the manufacture or use of 10 per cent of the 20 000 chemicals notified during this period. Also like Europe they have an extensive programme for testing and management of existing chemicals. It had targeted 17 high priority toxic chemicals for national reduction in release to all media by 33 per cent by the end of 1992 and 50 per cent by the end of 1995.

Unlike Europe, however, the EPA does not have a significant policy on chemical products although the law does regulate pesticides as products. There are few regulatory requirements for product registration and the USA has no product register. Perhaps the most significant legislation for chemical product registration is that enforceable under the Food and Environmental Protection Act in the UK. Other drawbacks of the EPA system are that although there is 'unified authority in the USA', registration takes place separately in every state. For instance, California, Virginia and Arizona have different data requirements. There are also problems with the TSCA and the Federal Insecticide, Fungicide and Rodenticide Act (FIFRA) which have different testing regimes for the same endpoint. Also the EPA system is based on risk assessment and exposure data which is evaluated before the data needs are defined.

8.8 Chemical Controls in the UK

As we have stated the UK has adopted all of the European Directives controlling release to the environment. Legislation in the UK, however, is unlike other countries in the EU and the UK has adopted its own set of regulations to control the release and exposure of chemicals in the environment. Before we consider the very special case of pesticide regulations in the UK it is worth explaining the differences of the UK system of law. Most of the laws brought about in the EU and to some extent the USA are based on Roman Law, as defined by the Treaty of Rome. The UK is different.

8.9   How Laws are Made

As we have stated laws in the EU are made in a different way to those in the UK. Indeed, each country has its own laws. In the UK there are many different laws but the laws concerning chemical control, although adopted by the UK, must pass through Parliament. When the UK joined the EEC in 1973 it agreed to obey all past and future EEC laws so that most Community law became UK law. European law, however, must be processed by the UK system before it becomes statutory and for this it must go through the due process of law.

All new laws must be agreed by Parliament and before this happens they must pass through a number of essential stages. First of all they are debated in the House of Commons and then in the House of Lords. Only after agreement by both Houses can they be passed on to the Queen who has to sign them to show that royal assent has been given. Then the law becomes an 'Act' of Parliament; before this it is known as a 'Bill'. Bills can start in the House of Commons or in the House of Lords and can pass through parliament in one of two ways as shown below.

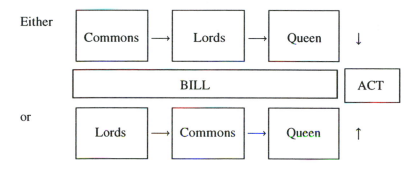

After House of Commons Parliamentary Education Sheet 1993 (*How Laws are Made*)

8.9.1   *Types of Bill*

There are three types of Bill: a Public Bill which affects everybody, a Private Bill which affects one area or organisation and Personal Bills which affect only one or two people. Most laws which control chemicals or pollution tend to affect all three areas. There are also different types of Public Bill. A Public Bill can be Government sponsored or brought about as a Private Member's Bill. The Private Member's Bill is usually brought by an individual MP or Lord. Historically, those brought by MPs seem to have more success than those originating from the Lords. All of the EU legislation which is to be adopted by the UK will be Government sponsored.

8.9.2 Consultation Stage

At the primary stages, the Bill will go through a process of consultation, drafting and approval: consultation between interested parties (e.g. the chemical industry), drafting to agree with government formats and approval by the Cabinet and its committees. The consultation stage is normally organised by the government department sponsoring the Bill, for instance for the Food and Environmental Protection Act and the Plant Protection Products Directive this would be the Ministry of Agriculture, Fisheries and Food (MAFF). At the drafting stage a 'green paper' is produced and after several drafts a 'white paper', which forms the basis of the Bill to be presented to Parliament, is produced. It should be remembered, however, that not all Bills require green and white papers. Civil servants are largely responsible for drafting Bills in the parliamentary council office at Whitehall. After drafting they can move on to the important third stage of approval. This involves the Bill being read out in the House of Commons.

8.9.3 First Reading

The first reading simply informs members that the Bill will be coming up for discussion at a future date. At this stage there is no debate in the House and Parliament does not need to know the content of the Bill. The Bill at this stage is often represented as a 'Dummy Bill' in the form of a single A4 sheet of paper.

8.9.4 Second Reading

At its second reading Parliament discusses the main purpose of the Bill. The minister in charge explains for the first time the purpose of the Bill. After this the Bill proceeds to the committee stage. Here all the details of the Bill are examined for the first time by a committee of MPs in a session outside general parliamentary debate.

8.9.5 Third Reading

The committees that look at the Bills are small, comprising about 18 MPs. The membership is proportional to the number of seats held by particular political parties; however, membership may also be influenced by the nature of the Bill. Only at the committee stage are amendments to the Bill allowed for the first time. The Bill can then proceed to the report stage, when Parliament is informed of the committee's actions and a reprinted version is presented to the House. This is then the third reading and the last stage before it enters the House of Lords. If the House is happy then it proceeds to the House of Lords.

8.9.6   *The Lords*

The purpose of going to the Lords is to get a second opinion. If the Lords rejects the Bill it is referred back to a first and second reading. The House of Lords is only allowed to delay a Bill for one year at which point it must be returned to the committee stage. The Lords has no ultimate veto over the Bill. The committee stage after the Lords is similar to that of the Commons but at this point the committee consists of the whole House and is held in the chamber of the House of Lords rather than the committee room. Any Lord who is interested can take part in discussions. A report stage then follows two weeks after the committee stage which allows the Upper House another chance to look at the Bill. After this the Bill proceeds to a third reading in the House of Lords after which, if there are no changes, it can be sent to the Queen for signing. If there are any changes it must be sent back to the House of Commons for agreement. The Commons normally agrees with the Lords' comments for non-controversial Bills. If there is disagreement which cannot be resolved the Commons will re-introduce the Bill the following year. In this case the Lords cannot reject a Bill which has been agreed by the Commons for two years on the run.

8.9.7   *Royal Assent*

Once the two Houses have agreed the Bill it is sent to the Queen for royal assent. The Queen then signs the Bill (she has little choice after Oliver Cromwell!) or gives assent when she signs what is known as a letters patent. This allows the Speakers to announce the results for the two Houses. After the announcement the Bill becomes an Act of Parliament and normally becomes law; however, to take effect it may need a commencement order. These are called 'Statutory Instruments'. This allows time for adaptations to the Act to take place. For instance this can allow the legislative machinery to be set up to bring into force EU Directives. The definition of a Statutory Instrument (1946) is *'Documents that grant delegated power'*. All Statutory Instruments are published by Her Majesty's Stationery Office and are on sale to everyone. Other papers include Command Papers, reports, etc. which are in effect edited white papers explaining specific Acts. As we have stated, many of the laws in this country are dictated by the EU and the laws on environmental safety cannot be considered in isolation. We discuss the way that the UK has adopted EU legislation shortly. There are notable exceptions to the way some countries control pesticides in the EU. An example is the UK scheme for the control of pesticides under the Food and Environmental Protection Act 1985.

The Food and Environmental Protection Act 1985 (FEPA) is defined as:

> An Act to authorise the making, in an emergency, of orders specifying activities which are to be prohibited as a precaution against the consumption of food rendered unsuitable for human consumption in consequence of an escape of substances; to replace the Dumping at Sea Act 1974 with provision for controlling the deposit of

substances and articles in the sea; to make provision for controlling the deposit of substances and articles under the sea-bed; to regulate pesticide substances or organisms prepared or used for the control of pests or for protection against pests; and for connected purposes.

16 July 1985.

The above Act is in three parts. Part I deals with the contamination of food and powers to make emergency orders, powers to ministers and powers of enforcement. Part II deals with dumping at sea and the requirements of licensing. Part III of the Act deals exclusively with pesticides and its scope is as follows:

(a)   With a view to the continuous development of means to

(i)   protect the health of human beings, creatures and plants.
(ii) safeguard the environment.
(iii) secure safe, efficient and humane methods of controlling pests.

(b)   With a view to making information about pesticides available to the public.

Under this Act, by regulation or order ministers have the power to control the import, sale, supply, use, storage and advertising of pesticides. Ministers also have powers to approve, review or revoke the use of pesticides and to make information available to the public. They impose data requirements, risk assessments and monitoring similar to the requirements under the new substances and existing substances regulations in the EU and adopted by the UK but often go further than these regulations. Ministers are empowered to impose codes of practice, charge fees and recover certain expenses. They also authorise enforcement officers to make sure that industry and individuals are complying with the Act. Parliament subsequently set up an enabling Act (Statutory Instrument) to allow this act to be enforced. The legislation was as follows.

8.10   Statutory Instruments under FEPA 1985

8.10.1   *No 1517. The Control Of Pesticides (Advisory Committee on Pesticides) Terms of Office Regulations 1985*

This Act came into force on 31 October 1985 when MAFF ministers and the Secretary of State acting under Section 16(7) of paragraphs 3 and 8 of schedule 5 of FEPA 1985a defined the terms of office for the Advisory Committee on Pesticides (ACP). It allowed for as many members as ministers saw appropriate to serve on an independent body of experts to advise on all aspects of pesticides including new active ingredients and reviews. A chairperson is appointed for a period not exceeding 3 years; however, in certain circumstances they may be appointed for a further term. The chairperson must sign a declaration that they have no financial or commercial interest in any of the proceedings. The ACP considers proposals on pesticides and advises ministers of departments that have signed the Control of Pesticides Regulations (COPR) 1986. These departments are the Ministry of Agriculture, Fisheries and Food (MAFF), the Department of Health (DoH), the Department of the Environment (DoE), the Welsh Office Agricultural Department (WOAD) and the Scottish Home and Health Department (SHHD). The ACP also delegates some of its work, especially the specialist scientific aspects, to a number of subpanels which report directly to the ACP. It has a Subcommittee on Pesticides (SCP) which considers the technical aspects of reviews and approvals, a Medical and Toxicological Panel and an Environmental Panel (EP) which deals specifically with the environmental aspects of pesticide approvals. There are, in addition, panels covering pesticide application technology, pesticide residues in food, and container and labelling design. The ACP takes notice of its panels but has the casting vote when recommendations are made to ministers. It considers and makes these recommendations after looking at the wider political and economical/commercial aspects as well as scientific implications. The main objective is to protect humans and the environment. As we have mentioned, there is also a great deal of other specialist legislation relating to the use of pesticides in addition to the FEPA. For agricultural pesticides the PPPD is now fully enshrined in UK law. The UK's obligations under this Directive run parallel to FEPA as very few agricultural pesticides have been entered onto annex 1 of the PPPD. This means that FEPA and the Control of Pesticides Regulations 1986 still apply to agricultural products. Furthermore, the Biocide Directive is not yet up and running, therefore non-agricultural pesticides are still legislated under UK law exclusively. This has created some administrative difficulties and recently the ACP has sought to rationalise and streamline its business and its panels. For instance, some panels now only meet on an *ad hoc* basis and the SCP has been renamed the Interdepartmental Sub-Committee and meets more regularly to handle the increasing volume of work from the EU. Work under COPR is still the main duty of the ACP.

8.11   Control of Pesticides Regulations (COPR) 1986

Figure 8.3 explains the general approval process for pesticides in the UK under COPR. COPR had its third reading on 3 July 1986 and became law on 6 October 1986. In the UK all pesticides must be authorised under this Act. These regulations cover both agricultural and non-agricultural pesticides, with MAFF being the authority for agricultural pesticides and the Health and Safety Executive (HSE) for non-agricultural pesticides. As part of these regulations MAFF and HSE act with the other government departments which are signatories to the ACP. In the UK non-agricultural pesticides include wood preservatives, insecticides, surface biocides and antifouling products. This is only a third of the likely scope of the proposed Biocides Directive; however, substances not covered by the FEPA are governed by the EU laws covering existing substances and new substances that we have discussed previously.

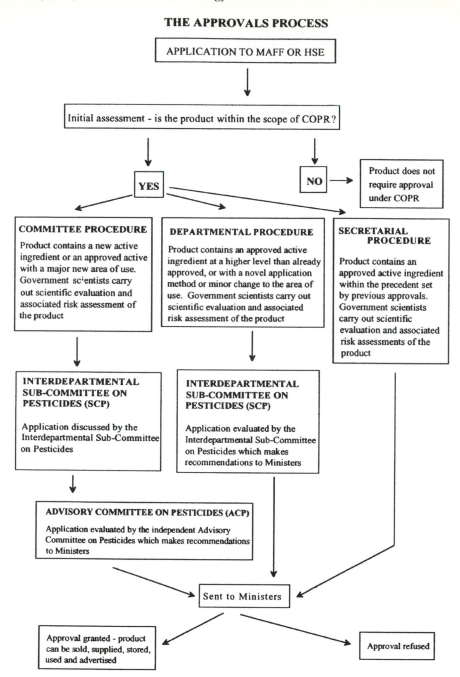

Figure 8.3 Approvals process for pesticides in the UK.

The purpose of the UK regulatory scheme is to have a system of law to deal with the authorisation of a pesticide and to have controls commensurate with risk and a scheme that is both efficient and cost effective. The system requires a company to supply a core set of studies which is roughly equivalent to the 'base set' package for new chemicals. Further data equivalent to level II and beyond depend upon the use patterns and the toxicity indicated by the core studies. Fate and behaviour form key elements of these further data requirements. Again like new substances an environmental risk assessment is carried out in order to predict the possible effects of exposure to the environment. Additionally, post-approval monitoring may be requested or experimental field studies can be carried out in advance if there is any concern for the effects. Eventually, after assessment, approval is given and a safety label applied. The cost of these studies is charged to industry. The cost is calculated according to the amount of time and effort taken to assess a package and as a proportion of the profits a company makes on an annual basis.

Applications from companies are dealt with either by the committee procedure or the registration department procedure according to guidelines laid down by the ACP. Through the auspices of the ACP, its various panels and the registration authorities, five ministers are involved in granting an approval to a company in the form of an experimental approval, a provisional approval or a full approval. The approval is the means by which ministers set specific conditions on the sale, supply, use and storage of individual pesticides. Ministers also set general conditions on the sale or supply of pesticides by publishing these on 6 October each year in the London and Edinburgh Gazettes. Anyone selling or supplying a pesticide is subject to these consents. These consents and the approval may be subject to a number of conditions and restrictions on the use of the pesticide and a minister can at any time review, revoke, suspend or amend an approval. All approvals are only granted for individual products and only for very specific uses. The three categories of approval are subject to varying levels of restrictions and data requirements.

8.11.1 *Automatic Approval under COPR*

An experimental permit can be given automatic approval to allow a pesticide use. This means it can be approved by a minister via the registration authority without passing through the ACP or any of its panels. This is only intended for research and development purposes and only provided (a) it is not handled by anyone other than researchers, (b) testing is carried out on site, (c) no product crops treated with the substance are used for human consumption or animal feed, and (d) all reasonable precautions are taken to protect people and the environment. Experimental permits not complying need to be assessed by the registration authority and passed to the scientific subcommittee in order to confirm their safety to humans and the environment and their efficacy. The trials are closely supervised, restrictions are imposed as appropriate and products and usage cannot be advertised.

8.11.2 *Provisional Approval under COPR*

Provisional approval is only granted for sale, supply, storage, use and advertisement of a pesticide if certain controls are met. All data requirements will have had to be met, an agreement on a safety label and text of data evaluation must be made available to the public. A provisional approval can proceed directly to a minister from the registration authority if a product contains an approved active ingredient within the precedent set by previous approvals. This is the most common route. Alternatively, a product may progress through all the departmental procedures. This can happen if a product contains an approved active ingredient at a higher level than already approved. Departments normally only pass through the SCP or other subpanels but independent government departments must agree on the recommendations after which a product gets approval for 3–5 years. An approval may need to pass through the full committee procedure if, for instance, a product contains a new active ingredient with a major new area of use. This means that however many subpanels the approval is assessed by it must pass through the ACP, be agreed by all departments and be signed by ministers before an approval is granted.

8.11.3 *Full Approval under COPR*

A product only gets full approval when all necessary data requirements and post-approval monitoring data are complete. Ministers must be satisfied that there is no long-term risk to the environment. Even then all approvals are subject to review at any time in the light of new evidence and as a matter of routine review.

The UK pesticide approval scheme is one of the most sophisticated in the world. FEPA not only enables the assessment of a product up to approval but sets regulations for monitoring humans and the environment for any adverse effects post-approval. With the introduction of FEPA, post-registration monitoring of pesticides became a mandatory condition of the approvals procedure. The surveys are funded from the pesticide levy. Pesticide usage survey groups established by MAFF for England and Wales and by the Scottish Agriculture Fisheries Department and by the Department of Agriculture Northern Ireland are coordinated by the working party on pesticide usage surveys. Their primary role is to provide the ACP with data and expert interpretation and technical advice on the usage of pesticides in agriculture, horticulture and food and grain storage. The scope of these groups is kept under constant review and there are often changes in the frequency and range of surveys. They do not, however, (as yet) cover non-agricultural pesticides. This is an area that will need to be addressed as in order to predict environmental concentrations of pesticides under the forthcoming Biocides Directive, approval of non-agricultural pesticides will be required to be accompanied by usage information.

8.12 Other UK Environmental Laws

8.12.1 *The Environmental Protection Act 1990*

Apart from the adopted EU laws there is a number of other UK Acts which protect the environment from pollution. As we have discussed before many of these have been adopted from EU Directives. The most significant Act to control pollution in the UK in recent years was the Environmental Protection Act 1990 (EPA). This Act's main purpose was to amend all other existing Acts regarding pollution and to bring them up to date as follows:

1. Make provision for improved control of pollution arising from certain industrial and other processes and to re-enact provisions of the Control of Pollution Act 1974 regarding waste disposal.
2. To provide extensions to Clean Air Acts 1951 (to prescribed gases).
3. To amend the Radioactive Substances Act 1960.
4. To make provision for genetically engineered organisms.
5. To make provision for the abolition of the Nature Conservancy Council (Wildlife and Countryside Act 1981) and create new councils (Welsh, Scottish, English).
6. To amend law relating to the Control of Hazardous Substances 1988 on, over and under land.
7. To amend section 107(6) of the Water Act 1989.
8. To amend sections 31(7)(a), 31A(2)c(i) and 32C7(9) of the Health and Safety at Work Act (HSW) 1974.
9. To amend the provisions of the Food and Environmental Protection Act 1985 as regards the dumping of waste at sea.
10. To make further provision with respect to prevention of oil pollution from ships.
11. Power to prevent burning of crop residues.
12. To make provision for financial or other assistance for purposes connected with the environment.

These 12 areas are enshrined in seven parts. Those marked with an asterisk are discussed in depth here.

Part i.   * Integrated pollution control.
Part ii.   * Waste on land.
Part iii. * Statutory nuisance and clean air.
Part iv.   Litter.
Part v.   Amendments to Radioactive Substances Act 1960.
Part vi.   * Genetically modified organisms.
Part vii.   Nature conservation in Great Britain and countryside matters in Wales.

These amend:

1.   The Countryside Act 1968.
2.   Local Government Act 1972 (Function of National Parks).
3.*   Wildlife and Countryside Act 1981.

plus 18 other Acts relating to pollution.

8.12.1.1 *Integrated Pollution Control*

The Secretary of State for the Environment defines pollution under these controls as 'any pollution of the environment with regard to harm to the health of living organisms or other interference with ecological systems from processes' prescribed or otherwise. A process is defined as 'any activity whether in premises or in a mobile plant which is capable of producing pollution'. A prescribed process refers to a process prescribed by the Secretary of State where an authorisation is required.

The Act allows the 'enforcing authority', in this case Her Majesty's Inspectorate of Pollution (HMIP) (or in Scotland, Her Majesty's Industrial Pollution Inspectorate and River Purification Authorities acting jointly), to issue an authorisation for a process which releases pollution into the air, water or land. In Northern Ireland this work is carried out by the Alkali and Radiochemical Inspectorate. This covers groundwater, rivers, lakes, the sea and the sea-bed. The EPA established the legal framework for integrated pollution control and brings together under the single control of HMIP discharges to air, water or land from the most polluting processes in the UK.

Processes resulting in pollution to air covered by HMIP embody the EU Framework Directives 84/360 for waste disposal and 88/609 for large combustion sites. The types of controls described here are air quality control standards and production standards for fuels and a framework for limiting emissions from industrial plants. There are particular air standards for sulphur dioxides, suspended particles, lead and nitrogen dioxide and production standards for fuel with a reduction in sulphur content and lead. Industrial plants are authorised to restrict the levels of sulphur, nitrogen, carbon monoxide, organics, heavy metals, dust, etc. (asbestos, glass, minerals) and chlorine/fluorine. Authorisations under the Framework Directive are based on best available technology not entailing excessive cost (BATNEEC) and interface directly with the EPA 1990. The EPA established the framework for integrated pollution control. For instance, more recently (1996) the Environmental Protection (Prescribed Processes and Substances Regulations 1991 SI.1991 No 472) prescribed that the BATNEEC must be used to prevent the release of triorganotins to the environment or if this is not practicable by such means BATNEEC must be used to limit the release of triorganotins and render them harmless. There are similar controls used to prevent or minimise the release of other 'Red List' substances to the aquatic environment. The Red List substances are those identified in Europe as being unacceptably toxic to the environment. Thus the controls will be aimed at minimising point source pollution from such industries as pesticide manufacturers and timber treatment works. HMIP are also responsible for collecting information on discharges from these processes and have power to enforce restrictions on emissions by fines or by closure of plants.

8.12.1.2 *Waste on Land*

The EPA 1990 additionally covers the pollution on land of any activity and embodies legislation laid down in EU Directive 86/2789. This Directive only protects soil when sewage sludge is used in agriculture and therefore goes further. Directive 86/2789 does, however, set limits on the concentrations of cadmium, copper, nickel, lead, zinc, mercury and chromium in soil following application of sewage sludge. The EPA also embodies the EC Directive on Hazardous Waste 91/989 and the Waste Framework Directive 91/156. These Directives control transfrontier shipment of hazardous chemicals and contain a list of hazardous wastes which must be controlled. The Directives aim to reduce the quantity and toxicity of wastes to landfill sites and to develop an adequate and integrated network of waste facilities with a view to prevention, recycling and reuse optimisation of final disposal, regulation of transport and remedial action. The EU has led to the policy of integrated pollution control. This is embodied in the Framework Directive which encourages member states to develop clean technologies and products which are designed to have minimal environmental impact by nature of their manufacture, use or final disposal.

8.12.1.3 *Statutory Nuisance and Clean Air*

We have already dealt with air pollution to some extent under the Integrated Pollution Control section. Again, general pollution of air is covered by Directive 84/360 and air standards are laid down for sulphur dioxide, suspended particulates, lead and nitrogen dioxide. Controls also extend to vehicle emissions. These include standards for carbon monoxide, nitrogen oxide and hydrocarbons. Here pollution is of a more diffuse nature and is policed by local authorities and vehicle registration centres.

8.12.1.4 *Genetically Modified Organisms*

The purpose of bringing genetically modified organisms under the EPA was to prevent or minimise damage to the environment that may arise from escape or release from human control of genetically modified organisms (GMOs). The Genetically Modified Organisms (Deliberate Release) Regulations 1992 provide for the environmental safety of releases of GMOs to the environment and marketing of GMO products. Regulatory control covering contained use of these products is covered by the Genetically Modified Organisms (Contained Use) Regulations 1992, which effectively implements two EU Directives.

8.13   Wildlife and Countryside Act 1981

This Act makes it illegal to intentionally kill, injure or disturb many different protected species. Doing so by proxy by polluting their habitat or poisoning directly any protected species is an offence under this Act. For instance, the general consensus is that all species of bat, both in the UK and Europe, are in decline or vulnerable. The 15 species of bat found in the UK are protected by the Wildlife and Countryside Act (WCA) 1981. The Act makes it illegal to intentionally kill, injure or take away any bat but it is also an offence to 'damage, destroy or obstruct access to any structure or place used by bats for shelter or protection, or to disturb a bat which is occupying a place or structure for shelter or protection' under this Act. Bats typically enter roof voids in domestic premises and spend at least part of their time in contact with roof timbers that have been treated with pesticides. Prosecutions have been brought under this Act for illegal poisoning of bats with organochlorine pesticides. Similar action has been taken against companies that have illegally allowed chemicals (dieldrin) to leach into the aquatic environment and kill herons. The HSE has now agreed to a universal classification and labelling scheme to protect bats from the effects of remedial timber treatment products as a result of this Act in combination with FEPA. As a result all remedial timber products intended for use in loft spaces must be labelled with the phrase:

> *Bats are an endangered species under the Wildlife and Countryside Act (WCA) 1981.*
> *Before using this product on or in any structure used by bats, English Nature, Scottish*
> *Natural Heritage or the Countryside Council for Wales must be consulted.*

8.14   The Water Resources Act (WRA), Water Industry Act (WIA) 1991

In the UK and Wales the above Acts, in conjunction with the EPA, provide the main framework for the control of water pollution and the implementation of water quality standards. They are mainly enforced by controls on direct discharges, pollution reduction technology and the establishment of environmental and drinking water standards. Similar legislation exists in Scotland and Northern Ireland. Under the above legislation the 10 major water companies were privatised and the National Rivers Authority (NRA) was set up to monitor pollution. The waters under its control include inland waters, freshwaters, freshwater rivers, canals, lakes, ponds, impounded reservoirs, groundwaters, coastal waters and territorial waters extending seawards for 3 miles. The NRA is committed under the WRA and various EC Directives (e.g. groundwater, drinking water) to monitor pollution by pesticides and chemicals, control their inputs to controlled waters and promote, like HMIP, the use of BATNEEC. Part III of WRA defines the means of controlling pollution by setting environmental quality standards and drinking water standards (EU). These are policed by the NRA and HMIP.

8.14.1 *The Drinking Water Directive 1980*

EU Directive 80/778/EEC was adopted on 15 July 1980 with a requirement that it be implemented by all member states. It relates to the quality of water intended for human consumption and has proved to be one of the most controversial Directives ever to come out of Europe. It sets a maximum permissible concentration of $0.5\,\mathrm{mg\,l}^{-1}$ for total pesticides in drinking water and $0.1\,\mathrm{mg\,l}^{-1}$ for individual pesticides regardless of their toxicity. Although the UK is signed up to this many regulators see these standards as a nonsense. They were not based on the hazardous properties of a chemical or the likelihood that chemicals would have an adverse effect on the environment. They were meant to represent a surrogate zero which at the time was based on the detection limit of lindane. Nowadays, we can detect down to nanograms of many substances. Equally, however, many substances found in the environment do not present a toxic profile at anything like the concentrations in the Directive. As a consequence the Directive is undergoing revision but it is questionable if any amendments will be made. It is likely that individual levels will stay the same with the level for total pesticides being removed. Nonetheless water companies in the UK are at present spending many millions of pounds of taxpayers' money to comply with this Directive.

8.14.2 *The Groundwater Directive 1980*

Directive 80/68/EEC has also been adopted by the UK and aims to protect groundwater against pollution by certain dangerous substances. It requires member states to prevent the introduction to groundwater of certain List I substances and to limit the introduction to groundwater of certain List II substances. List I substances include organochlorine pesticides, aldrin, dieldrin, isodrin, DDT, lindane and pentachlorophenol. List II substances are more general but largely cover all biocides and their derivatives used as pesticides. As yet after two resolutions and several reminders by the Commission community action on the groundwater Directive has not been forthcoming. Following a request from the Council of Ministers, the Commission in 1995 produced a plan to manage groundwater resources. Member States are currently attempting to follow this plan and only time will tell how successful these measures have been.

8.15 The Future

At the beginning of these discussions we outlined legislation focused on individual environmental media. Progressively we have seen a move towards a more integrated approach. The very latest views on legislation are that we should adopt a cradle to grave attitude in chemical scrutiny. There are many new incentives coming out of Europe at the moment. The Ecolabelling Regulations 880/92 encourage a voluntary scheme of positive labelling of products which do the least harm to the environment. Similarly, the Eco Audit Regulations COM (91)45 is a Commission proposal on environmental auditing. This aims to evaluate and improve the environmental performance of industrial activities and to provide information to the public. The results of this should soon be published. Similarly, the Toxic Release Inventory inspired by US legislation proposes a requirement by industry to publish an annual inventory of emissions from industrial plants. The UK EPA regulations actually cover this at present in the UK. The Biocide Directive will bring many additional chemicals under regulatory control throughout the EU. It may also bring in legislation whereby we can selectively compare one product with another and only allow safer or more efficient products onto the market. We continue to move towards a risk and cost benefit type approach to chemical legislation rather than merely looking at the hazards of a chemical. In 1990 the government produced a white paper 'This Common Inheritance'. This included many specific targets and objectives to control chemical substances in our environment. New commitments have been reported annually since then. These reports are seen as a practical expression of the bringing of sustainable developments to bear on the problems. Many other countries are also cooperating in order to 'relate the concept of sustainable development to their national policy making' and it is now a matter for international concern. This concern stemmed from the UN Conference on Environment and Development (the Earth Summit) held in Rio de Janeiro in 1992. The discussions did not primarily concern chemical pollution but they were part of an integrated approach to a global problem. Key areas linked to pollution were:

Agenda item 21

a comprehensive programme of action needed throughout the world to achieve a
   more sustainable pattern of development for the next century.

The Climate Change Convention

an agreement between countries establishing a framework for action to reduce the
   risk of global warming by reducing emissions of so-called 'greenhouse
   gases'.

The Biodiversity Convention

An agreement between countries about how to protect the diversity of species and
   habitats in the world.

The UK, in signing up to the Earth Summit, has published its own strategy for sustainable development. John Gummer's words at the beginning of this chapter highlight the need not to be complacent about pollution. Although the air quality in our towns and cities in the UK has improved dramatically over the past 30 years pollution from vehicle emissions is becoming a major problem. Acid rain caused by sulphur emissions from our power stations is still damaging our buildings and killing our forests. In the UK, although we have an abundance of water, growing demands have caused drought in many areas. Furthermore, a small proportion of our rivers still continues to be affected by severe industrial pollution. Pesticide levels exceed EU drinking water standards in several areas and pollution from nitrates and pesticides and from leaching from waste disposal sites and contaminated land continues to pose a potential risk to groundwater in some areas. Intensive agriculture continues to pose a problem to habitats and wildlife. There needs to be a concerted effort through legislation to reduce emissions and to find more benign chemicals to use as pesticides. Large quantities of waste are still being produced as an undesirable by-product of production. This threatens soil, sediment and water quality and the habitats and wildlife they support.

It is becoming increasingly clear that a new approach to the management of our environment is essential, however time is finite. Recent legislative initiatives have recognised the need for global awareness of the problems. Global awareness is one thing, global enforcement is another. It is no good if one-fifth of the world's population suddenly decides it has destroyed a significant part of the earth's ozone layer from the use of CFCs and decides it is time to stop if four-fifths of the world's population in China decides it is high time they acquired a fridge for the first time and start using these chemicals. Ultimately enforcement of environmental legislation must come about by international cooperation and embody integrated environmental management. If economics are allowed to overshadow the environmental impact we may be no further forward. Cost benefit analysis and BATNEEC are becoming the 'flavour' of the 1990s. In legislation, however, there must always be a place for informed scientific judgement. To reduce this to a small component in a bureaucratic juggling act policed by pseudoscientists, economists and the mandarins of policy would be a folly. Legislation should remain a symbiosis of like-minded people from different disciplines where all have their say: the industrialist, the scientist, the regulator and the conservationist. What should not be disputed is that the polluter must pay.

Further Reading

The best further reading for this chapter is to look up the specific Directives and Statutory Instruments in a law library.

9

Prologue to the Future

The chapter opens with Rachel Carson's 1963 controversial doom and gloom view of the future. It then covers examples of pollutants which are causing global effects:

- Chlorofluorocarbons and the ozone layer
- Packaging and the need to recycle
- Phosphate and eutrophication
- Sewage

The chapter ends optimistically. Carson's predictions were wrong, but made us think. Now is the time to act.

9.1 Life Since Rachel Carson's *Silent Spring*

It is nearly 35 years since Rachel Carson's *Silent Spring*. Carson's view of the future was extremely controversial because of the blame she placed on industry and agriculture for the environmental effects that were beginning to become obvious. To prevent her worst predictions coming true industry and farmers would have to significantly modify their approach to killing pests. Until Carson almost anything was acceptable, development and progress were all important. Very few people even gave a second thought to animals and plants and the important part that they play in maintaining the equilibrium of our planet. The kinder criticism of her work accused her of being an unscientific, emotional polemicist; others considered her an anti-establishment, anti-technological communist nut! Whatever anyone said, Carson had brought the environment to the forefront of public opinion and politicians' minds; she should be revered for that alone.

In truth, Rachel Carson deserved none of these reactive and emotive retorts directed by an industry under siege. She was in fact a very respected government scientist who worked for a number of years in the US Fish and Wildlife Service. Her work exposed her to the effects that post-war chemical products were having on the environment. Older chemicals were known to be extremely hazardous but effects in the past were relatively well contained. For instance, the application of lead arsenate in an orchard is very different from the blanket spraying by air of the defoliant Agent Orange in Vietnam in the 1960s. Most of the examples of dire consequences of pesticide usage cited in Rachel Carson's book were taken from the American experience. She was particularly interested in a series of events known as the 'Mississippi fish kills' in Louisiana in the early 1960s that resulted from the use of endrin in the sugar cane fields. Her condemnation of agricultural and chemical practices was far reaching. The military came under fire for manufacturing pesticides for their own use at a Rock Mountain arsenal near Denver. This resulted in unexplained sickness in livestock and extensive crop damage in the area. Elevated levels of DDE and toxaphene were found in fish and birds in the Rio Grande in Texas. Endrin used on grassland in Montana was having effects on migrating ducks. Heptachlor used to control the fire ant in Hawaii was found in high levels as a contaminant of milk. The list went on and on.

Rachel Carson was particularly concerned about chlorinated hydrocarbons and organophosphates. She saw these as the greatest threat to aquatic and terrestrial life. She presented evidence of bird and fish kills, of human nervous system disorders and deaths. Herbicides in agriculture were seen as particularly toxic. The effects of leaching, run off and direct spraying onto agricultural land resulted in the direct contamination of water courses. Water treatment works were powerless to detoxify these chemicals which interacted to form toxic broths. These she predicted could only increase in the future. She saw that chemicals in soils were leading to the destruction of beneficial organisms. Indeed the whole basis of food webs and energy transfer could be compromised to the point of extinction of all life. She thought that government scientists were blinkered and only concerned with the 'classical effects' of pesticides in laboratory animals and not the effects on wildlife or the accumulation in the environment. She also recognised the need to control residues in food. She considered that increasingly higher doses of pesticides would be necessary to control greater and greater pest problems because of the development of resistance by the target pest. DDT had brought with it the age of rampant resistance and, as discussed in previous chapters, the near demise of some birds of prey. A bleak picture indeed.

Fortunately not all of Rachel Carson's predictions have come true. Life as we know it has not ended. We must remember that her book was written over 30 years ago during the embryonic period of ecotoxicology. Furthermore it must be said that many of her assumptions were just assumptions. Many of her predictions were based on sketchy unsubstantiated data. What, however, her book was able to do was to clearly focus for the first time on the problems associated with the over-zealous and indiscriminate use of chemicals worldwide. The massive response by the public ensured that governments looked at the problem in a new legislative light and began to take the subject of ecotoxicology more seriously. The lack of scientific information on the effects of chemicals, particularly pesticides, on human health and the environment led to near hysteria in some sections of the population. Whatever the ethics of this situation it was not easy for governments to rationally and logically dismiss these fears without data or without the legislative framework to facilitate collection of these data.

What Rachel Carson achieved in the USA was to change the accepted view that rivers, lakes and the sea, because of their large volume, could cope with all the waste chemical products from agriculture and industry without environmental effects. Eloquently highlighting (however melodramatically) concern for the environment led to public pressure which in turn led to legislation and resulted in major reforms. Faced with a storm of public protest, the USA have, since the amendment of FIFRA (see Chapter 8) in 1970, introduced 30 pieces of environmental legislation aimed at controlling the effects of chemicals on people and the environment. As we discussed in Chapter 8 this has resulted in reciprocal regulation throughout the world and in particular in Europe.

9.2   Is Environmental Legislation Having an Effect?

Legislation has to be sensible and enforceable if it is to be effective. Much of the earlier legislation in the USA was enacted prematurely in response to strong public pressure. It was based on sketchy and subjective information and incomplete scientific data. Regulations in some cases were bureaucratic and at odds with different government agencies; indeed the public are still sceptical as to its real value. The popular and scientific press has done little, in some instances, to allay these fears. It must be said there have been many improvements over the past three decades. There is no doubt that in the USA the residues of many pesticides in the environment have fallen dramatically. By 1976 municipal and industrial waste water, river run-off, aerial fall-out and harbour discharges showed a reduction of some 95 per cent in the levels of DDT; levels of PCBs fell by a similar amount. DDT levels in fish and shellfish, although falling dramatically, have remained at 1974 levels. Furthermore, leaching of the more commonly used pesticides into groundwater has increased with the introduction of carbamate pesticides which are more water soluble. It should be said, however, that contamination has been at very low levels (i.e. in the ng dm^{-3} range).

In the UK similar problems have been experienced with the leaching of triazine (e.g. Atrazine) pesticides. Here contamination is more likely in agricultural areas that have high nitrate levels. The EU has recognised this problem and is seeking to reduce the levels in Europe by legislation enforceable under the Groundwater Directive 1992 (see Chapter 8).

Legislation is becoming more and more stringent, but we must not become complacent. Since Rachel Carson, the number of chemicals produced has increased rapidly to a point where the number currently on the EU market is about 100 000. Although many of these will be benign to the environment some have the potential to cause serious human and environmental effects. There are still too many persistent substances (often referred to as Red List chemicals) based on hydrophobic halogenated hydrocarbons or toxic metals that have the potential to build up to toxic levels in the environment and tissues. Pesticides, pharmaceuticals and veterinary medicines continue to be a problem as they are designed to be biologically active. A step in the right direction is that governments throughout the world (particularly in the USA and the EU) have now identified the most environmentally hazardous substances such as PCBs, heavy metals, persistent pesticides and ozone-depleting chemicals. Globally, governments are taking action to control and even ban their use. The process is slow but hopefully inexorable. For instance, we mentioned previously chemicals such as lead arsenate used in orchards. In the past few decades the use of lead-based compounds in agriculture has declined gradually and their use as anti-knock agents in petrol (a major source of environmental contamination) has declined very significantly due to public and government pressure on the petrochemical companies. Reduction in lead emissions and its use in pesticides, petrol, paints, plastics and batteries since 1974 has resulted in lead blood levels in children decreasing by 5 per cent per year. Tin, in tributyltin oxide (TBTO), used in antifouling paints had a devastating effect on the shellfish industry in the 1970s. The French oyster industry was all but destroyed. A ban on its use on small boats in Europe and the USA has resulted in a recovery of shellfish which were affected by shell deformations and atrophied musculature. Even though concentrations of TBTO in affected areas have decreased by 90 per cent since local bans in the 1980s, it cannot be said that all aquatic species have recovered. Indeed it is evident that certain species such as the common dogwhelk (*Nucella lapillus*) will never recover unless there is a universal ban of the TBTO products still used in antifouling paints on larger vessels. There is further evidence that many other antifouling products currently approved for use in the UK, USA and continental Europe may also be having an effect on the aquatic environment. The UK Government is currently reviewing these as a matter of urgency. That they are toxic to the environment is indisputable. What is required is to achieve a balance between the environmental effects and the benefits these products can bring, such as the saving on fuel and the effectiveness of coastal and ocean-going military craft in the defence of the realm.

9.3 Cost Benefit Considerations

The question of cost and benefit, although an important issue, should not cloud our vision as to whether the adverse effects of chemical pollution are increasing or diminishing. Without evidence of gross environmental effects it may be impossible to calculate the true cost in terms of ethical considerations. It is essential that certain fundamental processes such as photosynthesis are not inhibited. Furthermore, it is essential that biological processes are able to continue within their normal temperature range; increased environmental temperatures might lead to plants and poikilotherms being subjected to non-optimum operational temperatures. Fundamental to this is that human activities do not continue to promote the greenhouse effect or result in the destruction of the ozone layer which protects our planet from potentially lethal UV radiation. It is indeed worrying that the universal adoption of inert halogenated hydrocarbon propellants and coolants (e.g. chlorofluorocarbons, CFCs) in refrigeration systems together with an unchecked burning of fossil fuels have added to the greenhouse effect and the consequent reduction in primary production which underpins all life. There is also an increased risk of mutation and carcinogenic effects in plants, animals and humans due to higher levels of cosmic radiation (e.g. low wavelength UV) reaching the earth's surface. This is now beginning to manifest in humans as an increase in the incidence of skin cancer (e.g. melanoma, a malignant skin cancer which looks rather like a black naevus or mole).

9.4 The Ozone Layer

Rachel Carson in her predictions of doom did not predict the additional invisible and creeping threat to the destruction of the ozone layer from the use of halogenated hydrocarbons. Located between 20 and 50 km above the earth's surface, the ozone layer screens out approximately 99 per cent of potentially deadly UV radiation. As UV radiation reaching the surface of the earth increases, the number of skin cancers increases. As little as a 1 per cent reduction in upper atmospheric ozone has the potential to cause 15 000 new cases of skin cancer each year in a world population of 250 million. It also increases the number of cataracts, can cause reduction in crop yields and reduces phytoplankton production in the oceans. We now see, year in and year out, a dramatic increase in the incidence of skin cancer in the white races of the world (see Figure 9.1). The problem was first identified in 1974 when USA scientists warned that CFCs used as propellants in aerosols, in air-conditioning plants, as dry-cleaning solvents and in plastic foam used in fast food containers could destroy the ozone layer. The warning was not heeded and by 1985 the chemicals industry was producing 800 000 tonnes of CFCs annually. This was the same year that British scientists, experimenting with weather balloons in the Antarctic, discovered a massive ozone hole opening up over a large part of Antarctica. These ozone holes have increased over the past decade and are now present over the Arctic and some major conurbations throughout the world (see Figure 9.1). World governments were so worried about the discovery in 1985 that the Montreal protocol was signed by the USA, the EEC and 23 other countries in 1987. Its primary aim was to cut world CFC consumption by 20 per cent by 1994 and by another 30 per cent by 1999. The protocol came into force on 1 January 1989 and since then 100 countries have signed it and it has been revised twice.

The protocol now covers CFCs, halons, carbon tetrachloride, 1,1,1-trichloroethane, hydrobromofluorocarbons (HBFCs), hydrochlorofluorocarbons (HCFCs) and methyl bromide. In the EU new regulations were introduced in December 1993 to cut CFC use by 85 per cent by 1 January 1994 and to phase them out completely by 1 January 1995. There are similar conditions which have been imposed on the other chemicals mentioned above. The EU is well on its way to implementing these regulations. The UK will make sure that controls will fully comply with these regulations and will ensure that users of ozone-depleting substances move as quickly as possible to alternatives that do not deplete the ozone layer; however, it is recognised that developing countries will have more difficulty in implementing the protocol. They will be allowed a 10 year grace period before

The hole that spells cancer for children

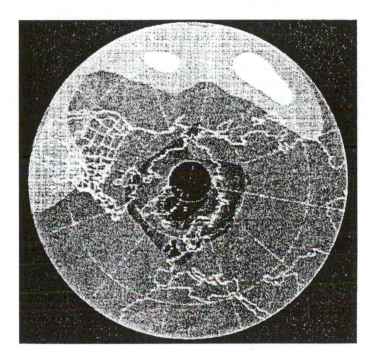

Figure 9.1 Depletion of the ozone layer over the Arctic region. The darker the monochrome the greater the depletion as reported by the *Independent* newspaper on 19 November 1996. Reproduced by kind permission of the *Independent*.

they are required to start following any of the phase-out schedules. As a result of this and because CFCs and other ozone-depleting substances remain in the atmosphere for a number of years it is likely that the situation will get worse rather than better over the next decade. Governments are universally worried.

The depletion of the ozone layer and the greenhouse effect are problems never predicted by Rachel Carson. Emissions of carbon dioxide and other greenhouse gases such as methane will probably result in rises in sea level of up to 2 m over the next 100 years. This is likely to have fundamental effects on agricultural practices and consequently living standards. The socioeconomic effects will be felt throughout the world. The UK is now committed under the FRAMEWORK convention on climate changes to returning emissions of carbon dioxide and other greenhouse gases to 1990 levels by the year 2000. To avert the rise in sea levels we may, however, be better placed to restrict the production of substances which deplete the ozone level.

9.5 Food and Drinks Packaging and its Recycling

9.5.1 *The Magnitude of the Problem*

There are a number of other issues concerning pollution where the EU and other countries are attempting to clean up society. In Rachel Carson's time the pace of the consumer society was only just in first gear. Food in shops was still being openly displayed and sold in paper bags. Supermarkets tended to package their products in paper cartons. Drinks were still sold in returnable bottles and the milkman delivered milk to the doorstep from the local farm. Today, if we consumed our weekly shopping immediately, we would be surrounded by a mountain of paper, plastic, metal and glass, most of which would end up at a municipal rubbish tip or landfill site. We all recognise the importance of packaging in protecting our food from damage and disease-causing organisms, keeping food hot in our fast food, fast living society and providing useful information on the nutritional value of the produce supplied. This packaging, however, creates mountains of waste that causes chemical pollution when dumped or incinerated and which results in acid rain and oestrogenic effects on fish, wildlife and possibly humans when the residues leach into water. Perhaps it is typical that the public are quick to recognise a problem, yet industry and government are still slow to take up the challenge, probably because of the potentially enormous financial impact that the necessary changes will have.

In 1987, the magazine *Good Housekeeping* carried out a survey which showed that the UK public rated bottle banks second only to lavatories in the facilities that they wished supermarkets to provide. Strangely most supermarkets still refuse to stock returnable containers but are active in providing bottle banks and facilities for disposal of used newspaper and aluminium cans. Unfortunately some only provide lip service to environmentally friendly disposal initiatives because there are no local recycling plants and bottles and cans simply end up with other domestic waste at council rubbish tips.

Disposal of newspaper is more of a success story as 30 per cent of the paper products we buy are made from recycled paper. That which is not recycled is at least biodegradable and breaks down rapidly in the environment by virtue of its cellulose structure. Recycling of metal cans can save up to 95 per cent of the energy used to make aluminium from scratch. This significantly cuts the air pollution which results from ore smelting.

9.5.2 *Plastic Packaging*

The problems of packaging produced from plastic are still a major problem. There are approximately 50 different types of plastics used in packaging, largely made from the by-products of oil, natural gas and coal. They represent about 7 per cent of the rubbish we dispose of, take a large amount of energy to produce from non-renewable fossil fuels, do not biodegrade and pollute the atmosphere when disposed of by incineration. There is growing evidence that very slow leaching of plastics containing phenol ethoxylates is responsible for the feminisation and sterility of fish stocks and possibly the reduction of the sperm count in western men. As the plastics are virtually indestructible under natural environmental conditions they are wholly inappropriate as components of disposable packaging. Most fresh food in supermarkets and food from fast food chains is packaged in polystyrene containers. There are very few plastic recycling schemes aimed at the public worldwide because of the sheer volume of plastics used in the packaging industry.

9.5.3 *The Need to Recycle Packaging Waste*

It has been estimated that 2.5 million tonnes of plastic, 6 million tonnes of glass bottles and 9.9 million tonnes of drink, food and pet food cans are collected as rubbish every year by local authorities. This compares with a total tonnage of 400 million tonnes of solid waste produced each year by industrial, demolition, construction and commercial waste. The trend is for an ever increasing amount of waste. With the ever increasing cost of waste disposal natural pressure is placed on industry to minimise and recycle waste. At the same time there is growing public awareness and governments (especially in the EU) are demanding higher and higher standards for protection of the environment. Current responses in the EU include a framework of UK and EU regulations to encourage the minimisation of waste while controlling its management. The UK Government's White Paper on the environment outlined a target of recycling half of all household waste that can be recycled (this is an unfortunate clause because it allows for very liberal interpretation of the White Paper) by the year 2000. This amounts to 25 per cent of all household waste. Similarly in the USA in 1990, the EPA introduced measures which required that all waste collection authorities produce recycling plans. With regard to packaging, there are plans for an EU Packaging and Waste Disposal Directive. To this end UK industries were asked to prepare a plan by December 1993 to introduce producer responsibility for packaging. These included plans for better organisation, a commitment to meet the costs of new collection and processing initiatives, a plan to recover levels of 50–70 per cent of packaging waste by the year 2000 and a plan to take immediate action to safeguard the recycling infrastructure for plastics and paper and board threatened by subsidised German imports. As a means of stimulating recycling the UK Government introduced a tax on landfill waste in September 1996 (see Figure 9.2). It will be interesting to see if this increases recycling rather than simply increasing the Government's tax revenue.

news

A cleaner way with rubbish: 'Eco' levy on landfill sites that

Tax on dumps promises jobs for thousands

NICHOLAS SCHOON
Environment Correspondent

Tomorrow sees the dawn of a revolutionary "eco" tax which will make polluting more expensive and create jobs. Yet, like any new tax, its introduction is surrounded by much moaning.

The pessimistic view of the Government's landfill tax – a levy on each ton of rubbish taken to a dump – is that it will lead to a surge of fly-tipping on roadsides, car parks and open spaces.

Furthermore, council-tax bills will rise or local government services be cut because the tax will be passed on to councils, which are among the biggest dumpers of all. In Ireland, rubbish will start to flow from Ulster to the Republic, where the tax does not apply.

But the optimists see the tax as a significant step towards an ecologically sustainable society. It will cause hundreds of new, labour-intensive recycling schemes to blossom, creating thousands of jobs. Some further jobs, though perhaps not many, will come from a small cut in National Insurance employer contributions which the tax is being used to fund.

This is the first application of a new taxation principle announced by the Chancellor, Kenneth Clarke, in his 1994 Budget. John Gummer, Secretary of State for the Environment, persuaded him to move towards more taxation of activities which do environmental harm (many of which are completely untaxed) and to reduce taxes on labour and employment correspondingly.

Landfill sites are environmentally destructive because rotting refuse produces methane.

of waste will be taxed £7, which drops to £2 a tonne for inert, non-rottable waste such as demolition rubble and ash. The tax will fall on the operators of landfill tips, who currently charge companies and local councils £5-£25 for each tonne of refuse received. The operators will pass the tax on to their customers.

It will bring in about £500m a year, and Mr Clarke has already pledged to use this to fund a 0.2 per cent cut in National Insurance employee contributions, taking them down to 10 per cent. HM Customs and Excise, which will collect the tax, estimates it will apply to around

His baby: John Gummer had pressed for a new approach

1,700 landfill sites. "We're not rash enough to claim that we have identified them all," a spokesman said. But Customs is fairly confident that the new tax will run smoothly because the sites already require a government licence and the amount entering has to be monitored to collect VAT.

new tax. Landfill site operators will be able to claim back 90 per cent of each pound of tax they pay in return for each pound they spend on approved "green schemes".

Those schemes will cover research and development into recycling and waste reduction, public education and the beautifying and greening of land blighted by disposal operations. They will have to be run by specially created, non-profit-making partnerships which can include tip operators, councils and environmental charities.

The new organisations will be controlled by a regulatory body which has not yet been set up. It is hoped that within a few years they will be spending tens of millions of pounds per annum, employing thousands of people in a range of schemes, many involving recycling.

The new tax will also promote the building of huge municipal incinerators which use the heat generated to produce electricity. Waste disposed in them is exempted from the landfill tax, so its advent makes them much more competitive with landfill sites.

It remains to be seen whether the tax, set at a modestly low level, will give the millions of households and companies who produce the waste an incentive to produce less. The latest figures show that in the South-East, the most affluent part of Britain, municipal refuse is rising by 3 per cent a year.

Mr Gummer hopes to persuade the Government to adopt other kinds of environmental tax linked to rebates for setting up trusts that run environmental improvement projects. "The

Figure 9.2 Headline from the *Independent* newspaper on 30 September 1996 heralding the UK tax on landfill waste. Reproduced by kind permission of the *Independent*.

Despite all the above legislation and worldwide government plans, any changes will be very gradual. Change, for example in Europe, will depend on the economic growth of the EU and the collective legislation that emerges from the European Commission and Parliament (see Chapter 8). Uncertainties about European law will inevitably affect the pace of change, but the trend will be towards waste reduction, reuse, recycling and recovery as opposed to the current trend of disposal through landfill. The UK Government is fully committed to reverse this trend. For instance, the UK has set up an Environmental Action Fund (EAF) which funds national and regional bodies to facilitate their development of schemes for reuse and recycling of waste. Also, since 1990, the UK's Department of Trade and Industry has given £4 million to recycling projects. This may seem small, but the Government's policy is to make the polluter pay. In 1993 the UK Government discussed with a range of industries, including packaging, newspaper, tyre and battery industries, ways in which they could recycle and recover their products when they become waste. Since the 1990s several industries have increased their recovery rates significantly. The largest recovery rate, for glass, plastics and steel cans, is 50 per cent by the year 2000; for aluminium cans the rate is 50 per cent by 1997–8 and for waste paper and board it is 40 per cent by the year 2000.

Industries are well on the way to reaching their target for paper and glass but for plastics, steel and aluminium cans only 5,12 and 16 per cent recycling had been achieved by 1992. For some time yet waste disposal through landfill will continue to account for the majority of UK waste, but better engineered sites and stricter legislation will result in less pollution to the environment. The trend for farmers to use paper by-products rather than rendered animal material to enrich soil is likely to continue, but stricter controls will be placed on the levels of leachates that are permissible in rivers. By the start of 1997, EU Directives will severely restrict the emissions from waste incineration plants. This will result in the closure of many, but it will reduce the release of dioxins into the atmosphere and may help to reduce acid rain. In the future the general public, with their acceptance of the need for sorting and recycling, will contribute to the reduction of environmental pollution. It is likely that products will have less packaging than in past eras and they may increasingly be required to return or reuse packaging. All of this should reduce the impact upon the environment.

It is encouraging to conclude this text on a positive note and there are two further areas where the consumer can take heart in the progress that has been made to date and the likely effects of past, current and future actions.

9.6 Phosphates and Eutrophication

Phosphates have been used extensively as agricultural fertilisers, washing powders and detergents for many years. They have also been responsible for extensive environmental pollution for many years and they were certainly a problem at the time of Rachel Carson. The main source of excessive phosphorus in the environment (apart from natural beneficial levels) was, and still is, from agricultural run-off and urban and rural sources derived from domestic sewage. Unlike the phosphorus derived from fertiliser from agricultural use, almost all the phosphate from urban and domestic sewage is from the use of washing powders and detergents. As phosphorus is lightly bound to soil particles the loss of phosphate by leaching is negligible. Input into fresh water is due mainly to soil erosion. Even so, the loss of phosphorus to water from agricultural land has been calculated as equivalent to 60 per cent of the fertiliser originally applied to the land. Much of the phosphate is tightly bound to soil particles and is less biologically available to plants. Phosphorus, mainly from washing powders, in domestic sewage remains an almost insurmountable problem.

The older type of detergents, the so-called hard detergents, are very persistent and contained high levels of phosphates; they were developed in the 1940s. Between 1950 and 1970 consumption of these detergents had increased fivefold in the USA and more than sevenfold in the UK. This paralleled the increase in prosperity after World War II. Many households were able to afford washing machines and the consumer moved away from the old soap powders based on simple fatty acids. As a result, phosphorus from these new detergents and washing powders amounted to 65 per cent of the phosphates in raw sewage by the early 1970s compared with about 20 per cent in 1957. The effects of phosphates and other nutrients in domestic sewage, which had increased in volume dramatically in many parts of the western world, was nothing short of disastrous. Its main effect was to cause eutrophication in lakes, rivers and streams.

Eutrophication is defined as 'enrichment of waters by inorganic nutrients'. Inorganic nutrients cause an increase in primary production which if unchecked creates algal blooms, decreases in species diversity, dominant species changes, increased turbidity (cloudiness) and sedimentation. During winter months, following excessive production in spring and summer, anaerobic conditions can prevail and anoxic conditions develop. Many lakes have in the past 30 years become devoid of most life except for simple bacteria. The Great Lakes on the USA/Canadian border suffered an ecological disaster in the 1960s and 1970s when phosphates from sewage outfalls significantly reduced the oxygen tension of water, resulting in the suffocation of fish and many other species. Although matters are improving several of the Great Lakes still have very low fish populations. A similar problem occurred in England when the quality of water in the Norfolk Broads on the eastern side of the country began to deteriorate in the late 1950s.

The Norfolk Broads are a group of small shallow lakes created in medieval times by flooding of old peat diggings. They overlie calcareous bedrock and were naturally eutrophic. The waters were clean and pure and supported a rich diversity of flora, fish and invertebrates. Many of the lakes were designated as areas of outstanding natural beauty and many were sites of special scientific interest (SSSIs). Unfortunately they also supported a vast and ever growing tourist industry based mainly on boating activities. This increased through the 1950s and 1960s. All of the normal household activities such as dishwashing, clothes washing and the use of toilet chemicals were carried out on pleasure craft. Unfortunately many of the chemicals used were simply dumped overboard. Similarly, domestic sewers containing high levels of phosphates also discharged into the Broads. The churning of propellers further increased the turbidity of the water by remobilising organic matter from the sediment and rapid eutrophication took place. A survey in 1972–73 of 28 Broads showed that 21 were devoid of any aquatic macrophytes or had extremely restricted growth. In general over the past three decades the Norfolk Broads as a whole have changed from a community supporting macrophytes and a rich diversity of fish and invertebrate life to one dominated by seasonal phytoplankton blooms. Organic enrichment also means they are slowly filling up with sediment, but there is light at the end of the tunnel.

In the USA up to 70 per cent of phosphates in sewage come from detergents. The elimination of phosphorus in detergents could remove 50 per cent of the total phosphorus entering lakes. Effective phosphate-free detergents have been developed and some states and cities in the USA have banned or restricted the use of phosphates in detergents. In Britain most detergents and washing powder manufacturers have now switched from the hard non-biodegradable-containing phosphates to more modern biological-type detergents.

9.6.1 Sewage Treatment

Most sewage works apply only primary settlement and secondary biological treatment to raw sewage. This at most removes 50 per cent of the nutrients. A tertiary treatment is necessary to remove the majority of phosphate in the sewage. Local authorities under pressure from government and the public are installing phosphate-stripping plants where sewage effluent outfalls to sensitive water bodies. The process is up to 95 per cent efficient. Such plants have been installed in the Norfolk Broads and on the Great Lakes in Canada. Recovery is not yet complete. An example is at Barton Broad in Norfolk, eastern England. Even though a phosphate-stripping plant was installed in 1981 and was, by 1984, reducing the load to the lake by 60 per cent, by 1984 aquatic plants were still not growing. In 1988 a study in America of 43 lakes that had received reduced phosphate loading showed that 44 per cent had not improved.

Studies have shown that integrated pollution control is necessary in addition to phosphate reduction in order to achieve full recovery. Governments in the USA and the EU are particularly concerned about eutrophication and the Paris Commission was set up to assess nutrients entering the North Sea and is seeking to reduce the levels of nutrients, including phosphates, reaching our rivers and seas. A UK research project (Ocean Interaction Study funded by the Natural Environmental Research Council) is reviewing the movement of nutrients, including phosphates, from the land to the North Sea. MAFF has also set up a research programme to investigate all major sources where phosphates are likely to enter aquatic water bodies from the land. It is evident that much more needs to be done. Populations grow and the volume of sewage increases year in, year out. Studies on rivers such as the Severn indicate that levels of phosphates doubled between 1974 and 1991. In the UK alone there are some 12 500 lakes, many of which may be threatened. It is interesting that one of our most picturesque and best loved lakes, Grasmere, contains dangerously low oxygen levels. These were at a depth of 10 m a decade ago but are now regularly found only 5 m below the surface. Phosphates may be playing a significant role in this phenomenon.

This chapter began with the doom and gloom of Carson's *Silent Spring* which frightened so many people and initiated our concern for the environment that in turn resulted in legislation in many countries to protect the environment. Pollution is global and we need to persuade the entire world to work together. Alas, politics makes this impossible. China continues to manufacture and use DDT as is evidenced by rabbit meat imports into the UK containing unacceptably high residues of DDT and its metabolite DDE. The UK exports its SO_2 and NO_x on the prevailing winds to northern Europe. Although clean air legislation will reduce the problem over the coming years only nuclear power would eliminate it, but we will not accept the risks, although they are very small, associated with nuclear reactors. Rain forests in South America and Indonesia are disappearing at an alarming rate; if this continues the earth will not be able to regenerate sufficient oxygen to sustain animal life. How do you persuade mahogany exporters, who stand to make millions of dollars from their environmentally unfriendly activity, that this might have dire consequences for their great-great-grandchildren? The problem is, of course, that they probably do not care.

Perhaps the most optimistic note to end on is that we are aware of the devastating effects that humanity can have on its environment and how it can reduce, or even stop, some of these effects. There is enormous public pressure to do something. Fortunately, if we act now it will not be too late.

Further Reading

Beaumont, P., 1993, *Pesticides, Policies and People*, London: The Pesticides Trust.
Carson, R., 1962, *Silent Spring*, London: Penguin Books.
Hoyle, W. and Karsa, D.R., 1997, *Chemical Aspects of Plastics Recycling*, London: Royal Society of Chemistry.

Index